Palgrave Studies in the History of Science and Technology

Designed to bridge the gap between the history of science and the history of technology, this series publishes the best new work by promising and accomplished authors in both areas. In particular, it offers historical perspectives on issues of current and ongoing concern, provides international and global perspectives on scientific issues, and encourages productive communication between historians and practicing scientists.

More information about this series at
http://www.springer.com/series/14581

Janet R. Bednarek

Airports, Cities, and the Jet Age

US Airports Since 1945

Janet R. Bednarek
University of Dayton
College Park, USA

Palgrave Studies in the History of Science and Technology
ISBN 978-3-319-31194-4 ISBN 978-3-319-31195-1 (eBook)
DOI 10.1007/978-3-319-31195-1

Library of Congress Control Number: 2016948762

Printed on acid-free paper

This Palgrave Macmillan imprint is published by Springer Nature
The registered company is Nature America Inc. New York

To M,
Love, J

Contents

List of Figures

List of Tables

Introduction: Airports, Cities, and the Jet Age

There are approximately 550 commercial airports in the USA. They range in size and level of activity from small, one-runway facilities with a few thousand aircraft operations per year to the massive Chicago and Atlanta airports with multiple runways and nearly a million such operations per year. Whether large or small, airports in the USA, for the most part, are locally owned and managed, and most originated as local responses to the advent of air travel in the 1920s and 1930s. These air transportation facilities—rooted in local history, local economy, and local culture—must also operate within a national and even global system of transportation, thus requiring often contentious dialogue and negotiation between various layers of government and between government and the airlines. The tensions produced, though present to a certain extent from the beginning, became acute after 1945 with the introduction of new technologies, jet airliners in particular; the shift from strict regulation of air commerce to a more market-oriented regime of deregulation; the phenomenal growth of air passenger traffic; the use of airports to promote local economic development, placing them within the long history of inter-urban competition; and broad security concerns that raised multiple constitutional issues. Most of these developments, often with unintended consequences or unfulfilled expectations, were beyond the full control of local airport owners, and all have combined to test the ability of cities to maintain, improve, and expand their airports. And they have even challenged the very definition of an airport. This book, then, is about how cities have

J.R. Bednarek, *Airports, Cities, and the Jet Age*,
DOI 10.1007/978-3-319-31195-1_1

managed their airports to meet the challenges of the rapid and profound changes in air travel since 1945 while, at the same time, hoping to have them serve as magnets for economic growth and development. In both endeavors, some cities have done better than others. In all cases, airports have changed the urban landscape.

The US commercial airport, like the airplane, was a creation of the twentieth century. Both were part of a technological revolution that in many ways transformed the way humans perceive time and space. Scholars examining transportation and mobility have focused particularly on the international airport as a space of global connection witnessing the continuous flow of goods and people. These mobility studies emphasize the role of international airports in a world marked by globalization and supermodernity, where the idea of both borders and security is challenged. These scholars view airports as key nodes in new systems of mobility, also including highways and computer networks, allowing for rapid, and in some cases even instantaneous, movement. And as parts of this global system, international airports have become, according to French social scientist Marc Auge, "non-places... a space which cannot be defined as relational, or historical, or concerned with identity."[1] In other words, because they must be easily navigated by those traveling within these systems of global mobility, international airports—somewhat like the shopping malls they have come to resemble—basically look the same the world over, reflecting not their locality, but the world of global travel.

Though social scientists have emphasized the "placelessness" of twenty-first-century airports, an examination of their longer-term development argues for their very rootedness in place. At least within the context of the political economy of the USA, the development of airports has reflected their local environments. The introduction of jet airliners, the explosive growth in the numbers of passengers traveling by air, deregulation, and security issues touched all airports, to a greater or lesser extent, between 1945 and the early twenty-first century. Decisions on how to respond to those broad changes rested with the local owners and managers. Often in contentious dialogue with federal officials, local airport managers negotiated solutions to the myriad issues they faced in maintaining—and perhaps enhancing—their city's place within a national and international air traffic system.

As a result of the expansion of the air travel passenger market, the introduction of jet airliners, and deregulation, cities have been continuously playing catch-up since World War II (WWII). While many cities came out of WWII with converted military bases that helped absorb the first new waves of air traffic, unexpectedly large growth in passenger traffic, new technologies, and changes in the rules of the game challenged even the best airport planning. Civic leaders expanded many airports with the expectation that the facilities could then meet demand for decades. Instead many of these same airports were overcrowded within a few years. On the other hand, some airport expansion projects predicated on expected passenger growth instead proved too ambitious in the wake of a collapse in passenger numbers due to the airline mergers, acquisitions, and bankruptcies common following deregulation. And as municipally owned properties, airports became logical places for cities to promote themselves to national and even international markets by adding or expanding runways and constructing distinctive terminals or other landmark structures. However, the local ownership and financing of airports have placed them at something of a disadvantage when competing with airports in other parts of the world where government officials clearly see them as national symbols and gateways and, hence, have provided extensive funding.

Additionally, strong negative reactions to jet aircraft noise have limited the ability of cities to expand their existing facilities or build new airports. Long before the term Not In My Backyard (NIMBY) was coined, airports became something no one seemed to want in their neighborhood. Further, aircraft noise became a distinctly local issue as court decisions made airport owners responsible for dealing with it. Both federal and airline officials called on local airport officials to use land use planning tools to deal with the noise issue. Though some efforts were made to shape development around airports, the general weakness of land use regulations forced most local airport officials to purchase and clear land to deal with noise complaints—no neighbors, no complaints. But even that did not always work as, for example, noise complaints increased after Denver opened its new airport in the middle of a nearly empty 50 square-mile clear zone

Finally, the issue of security has perhaps touched airports and individual travelers the most. Airports became the front line in the struggle against hijacking in the late 1960s and early 1970s. Security measures, however, remained rather minimal through the end of the twentieth century, balancing a need to make passengers feel safe with a desire to not inconvenience them. The September 11 terrorist attacks changed the security landscape at the airport. And while most of the attention focused on terrorists and hijackers, they were not the only targets of airport security measures. In the

1970s, air passengers began to complain about solicitations from various groups—many religious—at airports. Over time, airport officials did what they could to eliminate this form of passenger discontent. How much and what they could do hinged on the legal definition of an airport. Was it a public space or something else? In the end, although courts tended to define airports in ways that made it easier for officials to control or ban solicitations, the need to keep airports secure in the wake of September 11 provided the final justification for the actions taken.

Similar to what well-paved roads did for automobiles in the first decades of the twentieth century, early public airport construction helped transform the airplane from an entertaining novelty to a familiar form of transportation.[2] In that way, the history of American airport development dates to the early twentieth century.[3] This book, however, focuses on American airports since WWII during which time the often rapid and seemingly boundless expansion of air travel in the USA did much to transform the aviation and urban landscape.

Part One: From 30,000 Feet to Ground Level

The book begins with a chapter examining the major trends shaping overall airport development in the USA between 1945 and the early twenty-first century: the introduction of jet airliners, deregulation, air passenger traffic growth, economic development initiatives, and security. Jet airliners—designed and introduced in the 1950s—had a profound influence on the air traffic system in the USA and much of the world. To a great extent, they made fast and inexpensive travel possible. Their introduction, however, challenged local airport officials. Further, the first two decades of the jet age unfolded during a time when the US government tightly regulated commercial aviation. The regulations created a relatively stable, predictable environment for the airlines and their passengers. By the 1970s, however, consumer advocates and others began to argue that the stable and predictable environment came at a high cost to passengers. They advocated for deregulation of the nation's airlines to promote competition and lower ticket prices. The Airline Deregulation Act of 1978 ushered in a new, volatile environment.

Both before and after deregulation, the entire post-war period witnessed sustained, often explosive, growth in the number of Americans traveling by air. Cities across the country competed to capture as much of this air traffic as possible. Although most studies seem to suggest that air passenger traffic numbers generally reflected the existing urban hierarchy—with the largest cities capturing the highest numbers of air passengers and the fastest growing cities witnessing the most rapid passenger growth—many cities embarked on extensive airport development programs to lure as many travelers—business and leisure—as possible. This competition became particularly acute after deregulation as airport officials sought new ways to entice more passenger traffic in an era of crowded airplanes and long layovers between connecting flights. Local officials believed that strong growth in passenger traffic translated into or could be equated with strong economic growth. Therefore, as with sports stadiums and convention centers, many city boosters pinned their hopes for economic development on the success of their airports.[4] Finally, airport security became a major issue in the 1960s. As aircraft hijackings became distressingly common, federal officials, the airlines, and airport managers struggled to determine the best way to deal with the problem. Although hijackings were federal crimes and an international phenomenon, federal officials in the USA placed most of the responsibility for stopping hijacking with local airport managers and insisted on private, airline funding of security measures. Airport owners/managers and the airlines protested, but from the 1970s through 2001, airport security was essentially a local matter. The Federal Aviation Administration established security guidelines, but it was up to local officials and the airlines to finance and operate the system.

As noted, US airports are locally owned and managed, with cities, counties, and states as the most common owners. Increasingly since 1945, airports have also been turned over to independent public authorities in the name of management efficiency and easier financing, as public authorities frequently have the power to issue revenue bonds to cover expenses without the requirement of a public vote. Internationally, the post-war period saw a strong movement toward the privatization of public airports, with one of the most cited examples being Prime Minister Margaret Thatcher's privatization of the British Airport Authority (BAA), owner of a number of major commercial airports in England. Privatization has had its champions in the USA, but with few exceptions, the nation's commercial airports remain in public rather than private hands. And although most airport financing is local and private (airlines), federal funding has also been crucial.

Part Two: Response to the Jet Age

In the late 1950s and early 1960s, local airport managers, working both with and sometimes against the federal government, responded to the advent of jet airliners. But, regardless of whether civic leaders decided to build a new airport or expand on old one, airports needed to change to meet the needs of the jet age. What exactly would be needed and how to redesign airports to accommodate jet airliners, though, took a number of years to work out. During this time, many airport features travelers take for granted in the twenty-first century—from jet bridges, to moving sidewalks, to automated baggage carousels—were new and somewhat experimental. The introduction of sleek new airliners also inspired the design of new airport terminals that would reflect the speed, grace, and glamour of jet age travel. As trains gave way to airplanes, these terminals became the new "front doors" for US cities. As such, many civic leaders wanted their new airport terminals to function as iconic symbols for their cities. And, given the investments in airport expansion, mayors and others also hoped to reap an economic development boost. There has always been something of an "if you build it, they will come" rhetoric about airport projects. The latest manifestation of that is John Kasarda's "aerotropolis" concept—the idea that cities can attain global status building and developing their airports as international gateways.[5] To what extent airports could independently promote economic development, however, became and remains a subject of study and debate. There was considerably less debate over the value of adopting the "airport mall" concept in the 1990s.

Part Three: Jets, Noise, and Unhappy Neighbors

Part Three focuses on the issue of jet aircraft noise. Although aircraft noise proved problematic for airport neighbors almost from the beginning of commercial aviation in the USA, the issue took on extreme urgency following the introduction of the first generation of jet airliners in the USA in the late 1950s. Even though local airport officials and federal aviation organizations recognized that jet airliners were likely to provoke noise complaints, they were taken aback by the intensity and persistence of those complaints. Noise became and remains the single largest environmental issue facing commercial airports. The noise issue further illustrates the often strained relationship between the federal government and local governments over airports, especially after a crucial 1962 Supreme

Court decision placed full liability for aircraft noise on the airport owners and managers. Local officials argued repeatedly that the federal government—as well as the airlines and the aircraft industry—needed to take far greater responsibility for and make far greater efforts to deal with jet aircraft noise. For the most part, though, the federal government and others were perfectly happy to let local officials find the solutions, but within very restricted parameters. In the end, in many cases, airport officials dealt with noise by removing those making the complaints.

Part Four: Security: Hijackings, Hare Krishna, and September 11

Following the September 11 terrorist attacks on the USA, security measures have transformed the air travel experience. Whereas air travel was once about speed and freedom, it is now associated with long lines, full body scans, and extensive restrictions on what you can pack to carry on the airplane. Only ticketed passengers are allowed past security to the gate area so airport "goodbyes" and "hellos" must all happen in areas before the security checkpoints. The most stringent of these restrictions are associated with September 11. However, airport security concerns date back to at least the 1960s and, in many ways, the USA could so quickly implement new security measures after September 11 because the groundwork already had been laid. Before 1970, the idea that every passenger would have to pass a security check before boarding an aircraft would have been considered strange, if not in violation of the fourth amendment protection from unreasonable search and seizure. Repeated hijackings of US aircraft, however, resulted in the implementation and gradual expansion of security measures beginning in the 1970s. And both the court system and passengers generally came to accept these as necessary and lawful.

Fourth amendment rights were not the only ones to come under challenge at the airport. A number of religious groups viewed airports as prime locations for distributing literature and soliciting funds. When passengers complained, airport officials sought ways to restrict these groups. The religious organizations argued that such restrictions violated their first amendment rights to freedom of speech and religious expression. The resulting court cases gradually created a definition of an airport that allowed for greater restrictions on first amendment rights. The new definition also emphasized airports as places where security issues were paramount. By the early twenty-first century, airports were not simply transportation facilities—if they had ever been that in the first place—but security-focused, non-public spaces.

During the 1990s, I frequently drove to Cincinnati for the semi-monthly dinners of the Seminar in the City group meeting at the Cincinnati History Museum Center. At the time I had just started work on my first book on airport history and I had recently taken my first flight from the Greater Cincinnati/Northern Kentucky International Airport. My earlier work, focused on the period before 1945, also examined the issue of airport ownership. Sitting at a table with a number of urban historians from the Cincinnati area, I asked who owned the Cincinnati airport. No one could answer my question. To be fair, though, the airport that serves Cincinnati does have a unique ownership structure. Nonetheless, the history of US airports has been of peripheral concern to both urban and aviation historians. I hope that this book will highlight the importance of studying this vital part of the urban transportation infrastructure and landscape.

Notes

1. Marc Auge, *Non-Places: An Introduction to Supermodernity* (London: Verso, second English Language Translation, 2008), 63. For other examples of the recent social science scholarship examining international airports, see John Urry, *Mobilities* (Cambridge: Polity Press, 2007) and Mark B. Salter, ed., *Politics at the Airport* (Minneapolis: University of Minnesota Press, 2008).
2. For a discussion of the relationship between road building and the adoption of the automobile, see Clay McShane, *Down the Asphalt Path: The Automobile and the American City* (New York: Columbia University Press, 1994).
3. For an overview of the early history of airport development, see Janet R. Daly Bednarek, *America's Airports: Airfield Development, 1918–1948* (College Station: Texas A&M University Press, 2001); see also, John Zukowsky, ed., *Building for Air Travel: Architecture and Design for Commercial Aviation* (Munich and Chicago: Prestl and The Art Institute of Chicago, 1996) and Deborah G. Douglas, "Airports as Systems and Systems of Airports: Airports and Urban Development in America before World War II," in William Leary, ed., *From Airships to Airbus: The History of Civil and Commercial Aviation, Volume I: Infrastructure and Environment* (Washington, D.C.: Smithsonian

Institution Press, 1996). For histories of individual airports in the United States see Betsy Braden and Paul Hagen, *A Dream Takes Flight: Hartsfield Atlanta International Airport and Aviation in Atlanta* (Atlanta: Atlanta Historical Society; Athens and London: University of Georgia Press, 1989) and Daniel K. Bubb, *Landing in Las Vegas: Commercial Aviation and the Making of a Tourist City* (Reno and Las Vegas: University of Nevada Press, 2012). For a broad overview of post-war airport development see Alan Altshuler and David Luberoff, *Mega-Projects: The Changing Politics of Urban Public Investment* (Washington, D.C.: The Brookings Institution, 2003), 123–175.

4. For examples of the literature on economic development fads, see William G. Colclough, Lawrence A. Daellenbach, and Keith R. Sherony, "Estimating the Economic Impact of a Minor League Baseball Stadium," *Managerial and Decision Economics* 15 (Sept–Oct 1994): 497–502; Peter Groothuis, Bruce K. Johnson, and John C. Whitehead, "Public Funding of Professional Sports Stadiums: Public Choice or Civic Pride?" *Eastern Economic Journal* 30 (Fall 2004): 515–526; Roger G. Noll and Andrew Zimbalist, "Sports, Jobs, Taxes: Are New Stadiums Worth the Cost?" *The Brookings Review* 15 (Summer 1997): 35–39; David Swindell and Mark S. Rosentraub, "Who Benefits from the Presence of Professional Sports Teams? The Implications for Public Funding of Stadiums and Arenas," *Public Administration Review* 58 (Jan–Feb 1998): 11–20, and Heywood T. Sanders, *Convention Center Follies: Politics, Power, and Public Investment in American Cities* (Philadelphia: University of Pennsylvania Press, 2014).
5. For a discussion of pre-1945 airport enthusiasm, see Bednarek, *America's Airports*, 41–66; for the "aerotropolis" idea see John D. Kasarda and Greg Lindsay, *Aerotropolits: The Way We'll Live Next* (New York: Farrar, Straus and Giroux, 2011).

PART I

From 30,000 Feet to Ground Level

Chapter One: From 30,000 Feet: Airports and Aviation History Since 1945

The history of US airports since 1945 is highly complex. It involves the history of technology and the history of cities, as well as important parts of US economic, demographic, and political history. In looking particularly at the relationship between airports and cities in the post-war period, it is clear that two important events were of paramount significance: the introduction of jet airliners and the deregulation of the airline industry. Though jet airliners did not create the issue, they nonetheless became symbols of the most intransigent program facing local airport officials—the problem of aircraft noise. Their introduction also required significant updating of airport facilities including runways and terminals. Deregulation prompted the wide-spread adoption of the "hub-and-spoke" organization of airlines routes and a period of often chaotic competition between airlines. In response, cities hoped to take advantage of the more fluid situation to attract more flights and more airlines. The results were often mixed. In addition to those events, the entire post-war period witnessed a dramatic expansion in the number of Americans flying, debates over the role of airports in local economic development efforts, and an increased focus on airport security. Those trends also presented local airport managers with a host of challenges. What exactly those challenges were and how well local officials responded varied from place to place, but virtually no place with a commercial airport remained untouched.

J.R. Bednarek, *Airports, Cities, and the Jet Age*,
DOI 10.1007/978-3-319-31195-1_2

Introduction of Jet Airliners, the Planes That Roared

Not to overstate the obvious, the invention of the jet engine made the "jet age" of travel possible. Frank Whittle, a Royal Air Force (RAF) officer, in England and Hans von Ohain, a physicist who developed an interest in aircraft propulsion, in Germany developed simultaneously, but separately, the world's first jet aircraft engines. While at the RAF technical school at Cranwell in 1928, Whittle wrote a senior thesis on "Future Developments in Aircraft Design." In it, he explored two new possible power sources for aircraft—rocket engines or gas turbines to drive the propellers. Shortly thereafter, Whittle moved beyond the idea of having a turbine power a propeller to the idea of a pure turbine or turbojet engine. He filed a patent on this idea on January 16, 1930. RAF officials, however, saw little merit in his ideas and Whittle went on to other duties, including earning a degree in mechanical sciences in 1936 from Cambridge University. A year earlier, in Germany, Hans von Ohain earned a Ph.D. in physics, with minors in aerodynamics, aeromechanics, and mathematics. A flight in 1931 in a Junkers Trimoter had left von Ohain unimpressed with the plane's propulsion system. While studying for his degree, he developed his own ideas for an aircraft jet engine, which he patented in November 1935. German officials saw the potential value in the devise and classified it as "secret." Von Ohain then joined the Heinkel aircraft company where he immediately began work on a design of a jet engine. At the same time, Frank Whittle, encouraged by some of his professors at Cambridge, found private support for his jet engine idea. On April 12, 1937, Whittle successfully tested the world's first practical jet engine. Though Whittle built the first practical engine, von Ohain was the first to have his tested on an aircraft on August 27, 1939. Whittle's engine finally powered an aircraft on May 15, 1941.[1] Germany, England, and eventually the USA would all develop jet aircraft during World War II (WWII), but extensive application came only after the war.

While the military was the first to apply jet engines to aircraft, the civilian commercial sector soon followed. But just as the development of the jet engine happened outside the USA, so, too, did the first commercial aircraft applications. The British led the way with the deHavilland *Comet*, which entered service in 1952. By 1954, however, a series of accidents—including three mid-flight break-ups of the aircraft later traced to metal fatigue—resulted in the plane being grounded.[2] In the meantime, the Soviet Union launched the Tupolev Tu-104 in 1956. For the next two years, until a redesigned *Comet* entered service, the Tu-104 was the world's only commercial jet airliner.

The appearance of the British *Comet*, as well as advances in jet engines and the company's experience building large jet bombers, led the Boeing Corporation to explore the design of a commercial jet airliner. Boeing decided to finance the development of the new airplane—which came to be known as the Dash-80—as a prototype that would not only sell the idea of commercial jets to the airlines but also the idea of a jet tanker to the United States Air Force (USAF). The USAF signed a contract for a tanker version of the airplane—the KC-135 – in 1955. In the meantime, Boeing sought the first airline customer for the commercial version of the plane—the 707. Most US airlines showed little interest in the new jet as they had just purchased the first generation of post-war, piston-engine airliners, including the Lockheed Constellation and the Douglas DC-7. In ushering the commercial jet age in the USA, Juan Trippe and Pan Am took the lead. As it happened, by the mid-1950s, the Douglas Aircraft Company was also developing a commercial jet airliner, the DC-8, and Trippe began to play Boeing and Douglas against each other to get the most airplanes for the least amount of money. Pan Am's orders for 20 Boeing 707s and 25 Douglas DC-8s ushered in the US commercial jet age and other airlines soon followed. Boeing beat Douglas into service with the first 707s flying Pan Am's New York–London route in October 1958. The first domestic commercial jet flights began in December 1958.[3]

The advent of the 707 and DC-8 challenged cities hoping to brag of jet air service as it would require airport upgrades. For examples, both airplanes could carry more passengers than their piston-engine predecessors as the Constellation could carry around 100 passengers, depending on the configuration, while the early version of the 707 could carry 179. While more passengers might require larger gate areas, more importantly the new jets also required longer runways. The Constellation could safely take-off from a runway less than 6000 feet long. The 707, at maximum take-off weight, needed 10,200 feet of runway (though a later version required only 9000 feet) for safe operations.[4] In 2015, the longest runway at Chicago's Midway Airport—the busiest airport in the 1950s (see below)—is still only a little over 6500 feet long. Eventually, jet airliners would enter service that could operate safely from shorter runways, but that was well in the future in 1958.

The most vexing challenge, though, was the noise the first generation jet engines produced. Both the 707 and the DC-8 first flew with

Pratt & Whitney JT3C turbojet engines. In a turbojet engine, all the air from the compressor is sent to the combustion chamber, burned, and exhausted through the turbine to run the compressor and produce thrust. The exhaust exits the turbojet at supersonic speed. The mismatch between the speed of the exhaust (supersonic) and the design speed of the aircraft (subsonic) meant that the engine was not efficient in providing propulsion to the aircraft. It was also very loud. However, shortly after the appearance of turbojet engines, engine companies developed new bypass or turbofan engines. In a turbofan engine, some of the air from the compressor bypasses the combustion chamber and turbine. With less air passing through the combustion chamber and turbine, the turbine has to extract more heat energy from the hot air passing through it resulting in a colder, slower exhaust, better matching the design speed of the aircraft and increasing propulsive efficiency. The cooler, slower exhaust also meant that the engine was quieter.[5]

Rolls Royce developed the first turbofan or bypass engine, the Conway, in the 1950s. Though the Rolls Royce engine was used on a number of commercial airliners, both General Electric and Pratt &Whitney soon developed more widely adopted American versions of the turbofan engine. And while General Electric was first to develop an engine, Pratt & Whitney produced the first flight-ready engine, the JT3D, in July 1959. The JT3D was designed to be retrofitted on both the 707 and the DC-8. By the early 1960s, all US jet airliners used the quieter turbofans.[6] But quieter did not mean quiet. Cities and airport officials would push the aerospace industry to produce quieter engines from the 1960s onward. As a result, there has been significant improvement in engine noise, but the noise problem remains.[7]

Just as it had led the industry into the commercial jet age, Boeing once again led with the introduction of the "jumbo" jet in 1969. The idea of a very large airliner grew out of conditions at airports in the 1960s. As the number of passengers expanded rapidly (see below), many airports became highly congested—more passengers meant more airplanes. The Douglas Aircraft Company first proposed addressing the issue by increasing the seating capacity of its DC-8. Douglas added a 37-foot section to the DC-8's fuselage, increasing capacity from 189 to 259 passengers. The "stretched" DC-8-61 first flew in 1965. When Boeing looked to do something similar, engineers found that the wing sweep and landing gear location on the 707 would not allow for a simple "stretching" of the airliner. Instead, Boeing turned its attention to designing a new airplane.[8]

Boeing began design on the 747 with the experience it gained in the competition for a very large military transport. Boeing lost the competition to Lockheed, which went on to build the C-5 Galaxy. Boeing, though, built on what it had learned to very quickly develop plans for a new large passenger jet airliner. Pan Am, looking to build capacity on its international routes, once again served as Boeing's inaugural customer. In April 1966, Pan Am placed an order for twenty-five 747s, just months after Boeing had begun construction on the world's largest aircraft manufacturing facility, needed for the massive new jet. The first 747 rolled out of the factory in Everett, Washington, in late 1968 and the company began deliveries in 1969. Pan Am flew its first 747 on the New York-London route in January 1970.[9]

Like the 707 before it, the 747 revolutionized jet air travel. Depending on the configuration—one, two, or three classes of seating—the early versions of the new plane could seat from 366 to 550 passengers. And it could do so efficiently. The 747 had a lower cost per seat mile than any other jet airliner, lowering the cost of flying both domestically and internationally. It was especially efficient on long-haul international routes, where airlines could still make a profit while operating the plane at 50 % capacity. Later models of the 747 had a seat/mile cost less than one-third that of the 707. Further, the 747 ushered in the era of the wide-body jet as the McDonnell-Douglas DC-10 and Lockheed L1011 soon followed. And while both these planes, especially the L-1011, could operate off of shorter runways (8070 feet vs. up to 10,450 feet for the 747), the Boeing 747 captured and dominated the wide-body transport market into the early twenty-first century.[10]

While not all airports could accommodate the 747, mostly due to runway requirements, those that could gained some benefit from the advances that had been made in turbine engine technology. Even though it was much larger—the 747 had a fuselage cross section of 255.5 inches versus 148 for the 707 and weighed 836,000 pounds versus 336,000 pounds for the 707—its four new high-bypass engines reduced its engine noise. The 747 was not quiet, but it was quieter.[11] By the 1970s, though, battles over airport noise had dramatically slowed, if not completely stopped, the construction of new major airports in the USA. While building a new airport became extraordinarily difficult, the expansion of existing airports also increasingly met with vocal and determined public opposition. Technical improvements to airplane engines were not enough to quell the dissatisfaction of airport neighbors.

Commercial jets were getting larger and may have been quieter, but they were not getting faster. There were, though, many who dreamed of a different kind of jet airliner, one that could fly faster than the speed of sound. At the same time Boeing was developing the 747, it was also involved in a government-run competition to develop an American supersonic airliner. The US project started in the wake of British-French and Soviet programs to develop supersonic transports (SSTs). In June 1963, President John Kennedy established a joint Federal Aviation Agency (FAA)-NASA project to design an American SST. Even before the new airliner had made it past the drawing board, though, public opposition mounted. The proposed SST needed far more thrust than could be produced by existing turbofan engines. It would need four large turbojet engines and those engines were very loud. Not only would the plane's engines be loud, but when traveling at supersonic speed, the plane would produce a sonic boom. SST advocates hoped that the American public would come to accept sonic booms as the price of speed. Government testing in the 1960s, however, showed largely negative responses to repeated exposure to sonic booms. Preventing or subduing the sonic boom was just one of the technical problems the US program was never able to solve. The proposed US SST was also too small to operate economically. The combination of technical and economic issues led to the cancelation of the US program in 1971.[12] However, the failure of the US SST program did not bring an end to the controversy over supersonic jet airliners as communities near number of US airports, including those in the Washington, D.C. and New York City areas, witnessed strong public opposition to plans to fly the British-French Concorde on trans-Atlantic routes, centered on the noise issue.

Deregulation: The Triumphs and Trials of a Free Market

Until 1978, the Civil Aeronautics Board (CAB), a regulatory agency created in 1938 to protect what Congress believed an "infant" industry, tightly regulated airlines in the USA. The CAB had broad authority to regulate the entry and exit of airlines from air routes through the issuance of certificates of convenience and necessity (which allowed airlines to fly passengers on specific routes), to set airfares, to oversee the financial health of airlines, and generally to protect "certificated" airlines from competition. Though how the CAB fulfilled its mandate varied over time, it generally worked to restrict both the entry and exit of airlines from routes and to set fares at a level to guarantee a certain return on investment. It also carefully monitored mergers and acquisitions. The CAB saw its role as not just protecting the airline industry, but encouraging the development of an industry that would result in a truly national system of air transportation that would serve not only civilian passengers, but also be available in times of national defense emergencies.[13]

This tightly regulated system had consequences not only for the airlines but for the nation's airports as well. The CAB determined which airline or airlines served which city. It restricted competition on flights between the nation's largest cities, but also guaranteed flights to the nation's smaller cities. Cities wishing expanded air service had to woo not just the airlines, but the CAB as well. Cities threatened with diminished service could also plead their case to the CAB, as it could require an airline to serve specific cities. The CAB also ruled on the type of equipment an airline had to use to serve the various routes, contributing to the rapid switch from piston-engine aircraft to jets—and the consequent need for longer runways at many airports as cities had to convince the CAB that their airports could handle the new aircraft. Although both airlines and cities often complained, the CAB did create a very predictable, if not stress-free, environment.[14]

All that changed with the advent of deregulation in the 1970s. By the late 1960s, it became clear to many critics that the regulatory system no longer worked. They argued it protected weaker airlines, promoted inefficiencies, and kept airline ticket prices unnecessarily high. Economists, including Alfred Kahn, who would later serve as chair of the CAB, conducted studies that unfavorably compared the nation's major airline companies with unregulated, intra-state carriers in California and Texas. According to Kahn, these small airlines managed to operate with greater efficiency and lower costs than the regulated carriers. He and others began to argue for a significant overhaul of the airline regulation structure.[15]

The push to reform airline regulation came at the same time the regulation of other forms of transportation—railroads and trucking—was also under scrutiny. Transportation deregulation had strong support in the oval office as both Presidents Richard Nixon and Gerald Ford advocated for it as part of their general efforts to reduce the size of the federal government. In many ways, the deregulation fight over the airline industry took center stage. Congress held major hearings on the issue beginning in 1975. And once in office, President Jimmy Carter, a Democrat who had campaigned as a small-government economic conservative, viewed airline deregulation as an important precursor to the deregulation of the trucking and railroad industries as the lower airfares promised by deregulation might appeal to business travelers. Seeing the economic benefits of airline deregulation, these individuals might then endorse deregulation of other modes of transportation. Further, support came not just from congressional Republicans espousing a small government philosophy, but also congressional Democrats, such as Senator Edward Kennedy (D-MA), concerned with consumer issues. Initially, the focus was on reform of the existing system. Gradually, however, the emphasis shifted to the complete dismantling of the CAB, especially after President Carter appointed Albert Kahn chairman in 1977.[16]

While support for deregulation grew in Washington, the airlines—its unions, its pilots, and its management—initially bitterly opposed it. However, political momentum remained in favor of action. As early as 1975, President Ford appointed John Robson, investment banker and future head of the Export-Import Bank, as chair of the CAB. Under his leadership, the CAB began to experiment with expanded routes and fare competition, leading to Texas International's famous "peanut fare" followed by American Airline's first "Super Saver" fares. Such experiments accelerated once Albert Kahn took the reins. Faced with such CAB action even before formal legislative deregulation, many airline leaders, including Robert Crandall at American Airlines, decided it was better to have full deregulation rather than more of the CAB's partial lifting of certain regulations. Kahn, who initially had supported only broad regulatory reform, also came to the conclusion that Congress must lift all economic regulation of the airline industry. Congress passed the Airline Deregulation Act in October 1978. It called for a gradual phasing out of the CAB's regulatory functions, including route entry and exit, fares, and mergers and acquisitions, with the CAB scheduled to cease operations on December 31, 1985. The free market, not the CAB, would shape the airline industry in the USA.[17]

And the free market proved very challenging for the newly deregulated airlines industry as the first few years under the Airline Deregulation Act proved extraordinarily turbulent. In 1979, the USA witnessed a second oil shock. The first oil shock had happened in the 1973 when the Organization of Petroleum Exporting Countries (OPEC) embargoed oil sales to the USA in retaliation for US support of Israel in the Yom Kippur War. The second oil shock came in the wake of the Iranian revolution, which resulted in decreased global oil production. The price of a barrel of oil soared from approximately $15 to nearly $40 (approximately $129 in 2014 dollars), dramatically increasing the price of gasoline and jet fuel. Although oil prices began to decline as early as 1980, in July 1981, the US economy entered a deep recession that lasted 18 months. The recession was marked by high interest rates and high unemployment, the latter peaking at nearly 11 % just as the recession officially ended in November 1982. If those economic headwinds were not enough, in August 1981, the Professional Air Traffic Controllers Organization (PATCO) called a strike. As government employees—the air traffic controllers were employees of the FAA—it was illegal for them to do so, but nonetheless the organization had led strikes in the past that had resulted in better wages and working conditions. This time, however, President Ronald Reagan—a former union president—fired all striking controllers. He further refused to rehire any of the strikers. Instead, supervisory and military controllers stepped in until replacements could be hired and trained.[18] And all of this happened just as the airline industry was learning how to deal with a deregulated environment.

And deregulation happened at a pace far faster than the legislation technically had envisioned. Albert Kahn left the CAB when President Carter tapped him to lead the fight against inflation in 1978. His successor was attorney Marvin Cohen and under Cohen's leadership the CAB moved aggressively to implement the new legislation. For example, existing carriers immediately petitioned the CAB for new routes and within a month the CAB approved over 200 route extension requests from certificated airlines. And it soon contemplated allowing all qualified applicants to fly requested routes. Before deregulation, an airline had to prove that granting such permission was consistent with public convenience and necessity. The legislation had called for a gradual withdrawal of the need for airlines to prove public convenience and necessity, with that requirement scheduled to end January 1, 1982. With the idea of more immediately increasing competition, the CAB moved toward a position of immediately granting all qualified applicants certificates until it was proved that such an action was not consistent with public convenience and necessity. Small airlines protested, arguing that the CAB was moving too fast and that such a policy might allow the larger carriers to take over routes and drive them out of business. The CAB ignored the protests, instead supporting competition and market forces.[19]

The CAB not only allowed existing airlines to rapidly expand their route structure but also opened the way for new passenger airlines to enter the market. Under the legislation, the primary barrier to entry remained "fitness"—a CAB ruling that an airline was "fit, willing and able" to offer passenger airline service. The CAB retained the authority to determine fitness until the end of 1985. After that point, the responsibility shifted to the Department of Transportation (DOT). In the meantime, in line with its emphasis on increased competition, the CAB liberalized its procedures for determining the fitness of an airline. Critics argued that this would jeopardize safety, but the CAB countered that no airline could afford to operate in an unsafe manner. As with the case of awarding new routes, the CAB operated on the side of unlimited entry and maximum competition.[20]

As a result of deregulation and additional economic pressures, the 1980s witnessed a major shakeout in the airline industry. A few of the major carriers—American, United, Delta, and TWA, for example—developed the hub-and-spoke systems that have become a familiar part of air travel in the USA. Concentrating flights at large hub airports and feeding their systems from numerous smaller airports put these airlines in the strongest competitive position, with American proving the most successful. Some smaller carriers—Western, Republic, Allegheny, and Piedmont, as examples—sought to mirror the larger carriers' strategy, but with limited success as most ended up merging with other carriers. Another group of carriers followed a low-cost, low-fare strategy. These included new entrant People Express, as well as Continental, Braniff, and Frontier. The most successful of that group was Southwest, which had begun as an intra-state, low-cost, low-fare carrier and continued to profit and grow. Newer and smaller commuter airlines also sought to find their niche in smaller, low-density markets. Their need for access to the larger hub airports, though, led to partnerships with the major carriers. By the late 1980s, the eight largest carriers essentially controlled 48 of the 50 commuter airlines.[21]

Basically, some airlines found successful survival strategies and others failed. In 1978, the airline industry counted 24 major air carriers. By 1989, a series of mergers, acquisitions, and bankruptcies whittled that number down to eight major national carriers—American, Pan American, United, Texas Air Corporation (Continental), Delta, Northwest, TWA, and USAir. There were also two regional carriers, Midway (which had merged with Air Florida) and Southwest. Deregulation had spawned 18 new airlines but only 2 survived in 1989—Midway and American West, another regional carrier. Most of the new airlines went bankrupt and/or were acquired by other carriers.[22]

The rapid expansion of air routes and the waves of mergers, acquisitions, and bankruptcies not only had a profound effect on the airline industry, they also had a significant impact on the nation's airports. Local airport managers particularly saw opportunities related to one of the airlines' major survival strategies—seeking the economies brought through creating a hub-and-spoke route structure. Though many airports sought hub status, the major airlines, especially the largest (American, Delta, and United) tended to create and fortify their major hubs at what were already some of the nation's largest airports—Atlanta, Dallas-Fort Worth, Chicago, and Los Angeles, for example. However, by the early 1990s, the largest airlines often had multiple hubs. Delta, for example, had its primary hub in Atlanta, but also operated hubs in New York, Miami, and Dallas-Fort Worth, as well as Cincinnati and Salt Lake City. American's main hub was Dallas-Fort Worth, but it also had hubs in Chicago, Miami, and New York, as well as Raleigh-Durham and Nashville. United had its primary hub in Chicago, but also operated hubs in Washington D.C., Los Angeles, and San Francisco, as well as Denver and Seattle.[23] And, as noted, hubbing was not exclusive to the largest airlines. Piedmont established hubs at Charlotte, Baltimore, Raleigh-Durham, and Dayton, Ohio. Allegheny Airlines, soon to become USAir, had its major hub at Pittsburgh. And Northwest operated its major hub in Minneapolis, but also had a hub in Memphis.[24] So a number of cities enjoyed increased passenger flights at their airports as airlines built and expanded their hub-and-spoke systems.

But becoming a major airline hub often brought mixed blessings. For examples, hub status became a way for a city and its airport to enjoy increased visibility and air traffic. Expanding the airport to achieve hub status generally enjoyed the support of local business interests. However, local residents often objected fearing diminished property values and quality of life due to increased aircraft noise. And even though aircraft had become quieter since the early 1960s, they were still not quiet and increased flights meant increased noise.

Further, as the era of deregulation stretched into its second and third decades, the economic fall-out continued. Though aimed at increasing competition, deregulation in many ways limited competition as the number of major airlines continued to decrease and the ability of those same major airlines to dominate certain hubs grew. Both resulted in very uneven benefits as some areas of the country continued to witness competition enjoyed lower fares, while others with so-called "fortress hubs" saw increased ticket prices.[25]

And hub status could also be fleeting as the shake-out in the airline industry continued through the 1990s and into the early twenty-first century. At the same time airport managers in some cities like Atlanta and Chicago struggled to keep up with growth and demand, others in Cincinnati and St. Louis found themselves with problems of contraction. Both cities had served as hubs for major airlines—Delta in Cincinnati and TWA in St. Louis. Those airlines dominated service to those cities. The airport authorities operating those airports pushed aggressive expansion plans in the 1990s and early 2000s with an eye at maintaining and expanding their hub status. After TWA merged with American and Delta with Northwest, however, both airports lost their hub status. Instead of managing growth, the airport officials at Cincinnati and St. Louis officials were faced with the challenge of declining traffic.

If You Build It, Will They Come? Passenger Growth and Economic Development

Deregulation and the expansion of the number of airline hubs highlighted the complex relationship between cities, their airports, and the airlines serving them. The relationship between cities and various transportation technologies—particularly in terms of competition for urban growth and economic development—has long been studied.[26] It has been clear that over time civic leaders invested heavily in new transportation technologies with the goal of enhancing their city's position in the urban hierarchy and capturing as much of the business of the surrounding hinterland as possible. The first large cities in the USA—New York, Boston, and Philadelphia—all relied upon their position as seaports for their livelihood. New York emerged as the clear leader in the early national period in part due to investment in fast sailing ships operating on fixed schedules. As the new USA reached into the interior of the continent, ambitious civic leaders latched onto canal and, by the 1830s, railroad schemes to promote their city's growth and development. Though Baltimore's boosters ranked among the first to invest in railroads, the leadership in the young city of Chicago soon capitalized upon that city's position as a potential intersection between rail and water to emerge as the nation's "second city." While Los Angeles is closely associated with the automobile, its emergence as a great city in the twentieth century also grew from efforts by its civic leadership to not only take advantage of its climate but also to defy nature and establish for the city a great seaport where none naturally existed. With the advent of the airplane, any number of cities hoped to capitalize on the new form of transportation. The long-term economic development impact (realized or potential) of airports for cities has not been extensively studied as yet. Those few existing studies have raised more questions than they have answered.

The Air Age, Growth of Passenger Traffic, and the US Urban Hierarchy

One way in which cities have used their airports to promote economic development has been through demonstrating strong and expanding numbers of passengers using their airport. As the number of airline passengers expanded tremendously in the post-war decades, a majority of airports could brag of increased passenger numbers (Table 1).

Although the chart might suggest uninterrupted growth until after 2005, there were several years in which passenger numbers declined over the previous year—1969–1970, 1974–1975, 1979–1981, 1988–1989, 1990–1991, 2000–2002, and 2007–2009. The downturn in passenger numbers generally coincided with downturns in the US economy, though other factors also contributed. The number of passengers in the USA peaked in 2007 at nearly 7.7 million.[27] Nonetheless, the growth in passenger traffic

Table 1 Passenger traffic, 1950–2010 (x000)

1950	19,220
1955	41,709
1960	62,257
1965	102,920
1970	169,922
1975	205,060
1980	296,901
1985	382,023
1990	465,558
1995	547,775
2000	666,149
2005	738,628
2010	720,496[a]

[a]See "Annual Results US Airlines" http://airlines.org/data/annual-results-u-s-airlines-2 (accessed June 20, 2016)

exceeded the overall growth in the US population. Between 1950 and 2010, the US population a little more than doubled from 150.69 million to 309.35 million. During the same time period, the number of passengers on US airlines soared from 14.541 million to 743.098 million—a factor of 51.

Changes in record keeping over the years makes it difficult to compare airport passenger traffic over the entire post-war period. Prior to the 1960s, published lists of the "busiest airports" often focused on aircraft movements rather than passenger totals, for example. Starting in the 1960s, though, passenger traffic per airport is available. Examining the list of busiest airports at several points in time for which statistics are readily available—1954, 1966, 1976, 1985, 1995, 2005 and 2010—reveals both remarkable continuity, but also a long-term shift in the location of the busiest airports away from the North and East to the South and West, mirroring the shift of US population during post-war period. It also reveals the advantages hub airports held in the hub-and-spoke system that emerged after deregulation (Table 2).

In 1954, Chicago Midway held the title of "busiest airport in the world." That title was based on aircraft movements, but as those movements largely involved commercial aircraft likely there was a close correlation between aircraft movements and passenger totals. The other nine "busiest airports" were New York-LaGuardia, Washington National, Los Angeles, Dallas Love, San Francisco, Cleveland, Miami, St. Louis, and Atlanta. By 1966—based on passenger data—the list looked much the same. Chicago was still on top, though the airport was the newer, larger O'Hare, not Midway and New York-Kennedy (new to the list) replaced New York-LaGuardia in the second spot, though LaGuardia remained in the top ten. Otherwise, while there were some shifts in the rankings, the list remained remarkably consistent with Los Angeles, San Francisco, Atlanta, Washington National, Miami, Dallas-Love, and New York-LaGuardia remaining. Missing were Cleveland and St. Louis, both of which would experience long, local battles over airport expansion plans beginning in the 1960s. They were replaced by New York-Kennedy and Boston. And as late as 1976, the list remained quite stable. Eight of ten remained on the list, though the Dallas airport was the new Dallas-Fort Worth facility, rather than Dallas Love. Washington National and Boston Logan had fallen to 11th and 12th places, respectively. Replacing those two cities were Denver, a western city, and Honolulu, a resort/tourist location.[28]

Table 2 Ten busiest US Airports by year

Rank	*1954*	*1966*	*1976*	*1985*	*1995*	*2005*	*2010*
1	Chicago Midway	Chicago O'Hare	Chicago O'Hare	Chicago O'Hare	Chicago O'Hare	Atlanta	Atlanta
2	New York LaGuardia	New York Kennedy	Atlanta	Atlanta	Atlanta	Chicago O'Hare	Chicago O'Hare
3		Los Angeles	Los Angeles	Dallas Fort Worth	Dallas Fort Worth	Los Angeles	Los Angeles
4	Washington National	San Francisco	New York Kennedy	Los Angeles		Dallas Fort Worth	Dallas Fort Worth
5	Los Angeles	Atlanta	San Francisco	Newark	Los Angeles	Las Vegas	Denver
6	Dallas Love Field	Washington National	Dallas Fort Worth	Denver	San Francisco	Denver	New York Kennedy
7	San Francisco	Miami	New York LaGuardia	San Francisco	Denver	New York Kennedy	Houston
8	Cleveland	Dallas Love Field	Denver	New York Kennedy	Phoenix	Phoenix	Las Vegas
9	Miami	New York LaGuardia	Miami	New York LaGuardia	Detroit	Houston	San Francisco
10	St. Louis Atlanta	Boston	Honolulu	St. Louis	St. Louis Las Vegas	Minneapolis-St. Paul	Phoenix

Yearly rankings based on total aircraft movements (1954 only) and passengers (all other years)

In looking at the lists between 1954 and 1976, though the changes were not dramatic, there were winners and losers—airports that had and would continue to rise or continue to fall in the rankings. The biggest upward movement between 1954 and 1976 involved Atlanta. Its airport went from the tenth spot to the second spot. And it would continue to grow, becoming the nation's busiest airport in 1995. Denver's airport (first Stapleton and then Denver International) first made the list in 1976 and then remained in the top ten through 2010. On the other hand, both Boston and Cleveland, which dropped off the list by 1966, never regained top ten status. Changes became more visible after deregulation.

The immediate effects of deregulation, especially the creation of airline hub-and-spoke systems, were evident as early as the 1985 list. Although Chicago, Atlanta, Dallas Fort Worth, and Los Angeles retained their top four rankings, there were many changes within the rest of the list. Miami and Honolulu just slipped below the top 10 ranking at 12 and 13 respectively. Replacing them were not Sunbelt cities, but Newark and St. Louis, both of which became important hub airports in the early 1980s.[29]

Deregulation and the creation of new hubs was not the only cause of change, however. The long-term shift of the US population to the Sunbelt and its effect on air travel also grew even more apparent after 1985. Once again, at the very top, there was remarkable continuity: Chicago, Atlanta, Dallas/Fort Worth, and Los Angeles, remained the busiest airports. The only change came in 1995 when Atlanta changed places with Chicago, with the former rising to the rank of number one and the later falling to number two, and Los Angeles and Dallas/Fort Worth, with the former ranking number three and the later number four. However, after 1985, the rest of the list was fairly dynamic with several cities making impressive gains and others seeing substantial slips in the rankings. The big gainers were all Sunbelt cities: Phoenix, Houston, and Las Vegas. Phoenix and Las Vegas, important resort destinations, made the most impressive gains. Phoenix jumped from 17th in 1985 to the 7th spot in 1995, remaining in the top 10 through 2010. Las Vegas started at number 24 in 1985, leaped to number 10 in 1995, and has remained in the top 10 since, climbing as high as number 5 in 2005 before falling to number 8 in 2010. Houston also made it into the top 10, but at a slower pace, rising from number 18 in 1985, to 14 in 1995, to 9 in 2005 and then 7 in 2010.[30]

The more fluid deregulated environment, rather than the shift to the Sunbelt, however, was the primary cause behind the airport experiencing the largest loss. As noted, St. Louis, which became an important TWA hub in the 1980s, had made its way back into the top ten, ranking as the tenth busiest airport in 1985, improving to ninth in 1995. However, once American Airlines purchased TWA following the latter's bankruptcy and closed the hub, St. Louis's airport no longer ranked in the top 10 or even the top 25.

As detailed below, a number of scholars who have examined the relationship between cities and airports have found that airport growth generally reflects the existing urban hierarchy, and thus also reflects broader continuities and changes within that hierarchy over time, such as the dominance of New York City and the continued high rankings of cities such as Chicago, San Francisco, and Philadelphia as well as the shift of population to the South and the West. For example, New York City has held to top spot on the list of largest American cities since the first census in 1790. Individually none of its three major airports topped the list of busiest airports in 2010, but if you combine the passenger traffic from all three (JFK, LaGuardia, and Newark), New York City ranks as the number one passenger destination in the USA in 2010 with 51,507,302 passengers compared to 43,130,585 in Atlanta.[31] Further, 14 of the 25 largest cities in the USA in 1950 have 16 of the 25 busiest airports in 2010 (New York and Washington, D.C. each have two airports in the top 25): New York, Chicago, Philadelphia, Los Angeles, Detroit, Baltimore, Washington, D.C., Boston, San Francisco, Houston, Seattle, Newark, Dallas, and Denver. Of those cities, all but Baltimore and Newark still rank in the top 25 largest cities, though Baltimore is close at number 26. New York, Los Angeles, Chicago, Dallas, and San Francisco had top 10 airports in both 1954 and 2010. Most of the cities having top airports in 2010 that were not among the largest cities in 1950 are Sunbelt (Southern and Western) and/or resort cities: Las Vegas, Phoenix, Charlotte, Miami, Orlando, and Fort Lauderdale.

The influence of the shift to the South and the West is also clear when looking at metro areas that grew in the post-war period and in 2010 have some of the busiest airports in the USA. For example, cities that did not rank in the top 25 largest cities in 1950, but do have some of the busiest airports in 2010 include Atlanta, Minneapolis, Charlotte, Miami, Salt Lake City—all but Minneapolis being Southern and Western cities. Many of those Southern and Western cities still do not rank in the top 25 largest cities, but their sprawling metro areas are among the most populous. Atlanta, which has had one of the busiest airports since the 1950s and has had the top ranked airport since the mid-1990s, was the 33rd largest city in 1950 and actually fell to number 40 in 2010. However, its metropolitan area ranks ninth in 2010. Miami is even smaller than Atlanta, ranking at number 44 in 2010, but its metro area ranks just above Atlanta's, at number 8. Charlotte, North Carolina, actually entered the top 25 largest cities only in 2010, though its airport has ranked in the top 25 since 1985, in this case also benefiting from deregulation as it served first as a Piedmont hub and then as a major hub for USAir. Only Salt Lake City falls below the top 25 as both a city and a metro area.

Conversely, cities that were among the most populous in 1950 and both no longer rank among the top 25 cities nor have among the busiest airports in 2010 include St. Louis, Pittsburgh, Cleveland, Milwaukee, Buffalo, New Orleans, Kansas City, Indianapolis, and San Antonio. With the exceptions of New Orleans and San Antonio, none of those cities are in the Sunbelt. As noted, the loss of hub status undermined the fortune of the St. Louis airport. That was also the case with Pittsburgh. Unlike St. Louis, Pittsburgh's airport never ranked in the top 10, but as the hub for Allegheny Air and later USAir, Pittsburgh ranked in the top 25 through the first couple of decades of deregulation. By 2005, after USAir decided to shift operations to its other hubs, the Pittsburgh airport fell to 38th place. It fell further to 46 by 2010. Cleveland's airport also lost ground following the 2010 merger of United and Continental and faced a severe restriction of operations once United announced it would close the Cleveland hub in 2014. There are similar stories in Milwaukee and Kansas City, while other airports, such as Indianapolis never operated as a major airline hub.[32] With both winners and losers amidst a good deal of continuity, the exact role airports play in shaping the urban hierarchy, including any independent role in the promotion of economic growth and development, remains a subject of booster dreams and scholarly debate.

The Air Age and Economic Development

As localities sought to meet the challenges of first the jet age and then deregulation, many advocates justified airport expansion plans on the premise, to paraphrase, that "if you build it, they will come." Airport expansion would result, they argued, in more jobs and enhanced economic development not only in nearby areas but in the entire metropolitan area. But the link between airport development and economic development at the metropolitan level has become a subject of great debate. While some studies identified a positive relationship between growth in passenger traffic and healthy metropolitan area population and job growth, others reached less bullish conclusions. Just as with other infrastructure projects that promised additional jobs and economic development—from convention centers to sports stadiums to aquariums—the performance of airport development projects varied.[33] For some, but not all cities, such investments paid rich dividends. In all cases, however, the question remained whether airport extension projects independently stimulated metropolitan economic growth or whether they served merely as one part of a much more complicated set of factors involved in economic development.

Serious study of the airport-economic development relationship dated back to at least the 1950s. In a pioneering 1956 study, urban geographer Edward J. Taaffe pointed out that a few of the nation's largest cities, including New York, Chicago, and Los Angeles, garnered more air traffic than other large cities, and that some large cities not only lagged behind but also attracted less air traffic than their population would have predicted. Taaffe demonstrated that large cities within 120 miles of dominant air traffic centers generated less air traffic because the dominant centers, such as New York City, Chicago, or Atlanta, absorbed more than their share from the perspective of comparative populations. Those places suffered from what he called the traffic-shadow effect. Examples included Milwaukee in the traffic shadow of Chicago and Birmingham in the traffic shadow of Atlanta. On the other hand, cities experiencing higher air traffic than their populations suggested shared certain characteristics that exempted them from the shadow effect. Some—like Denver and Salt Lake City—lay outside the range of overnight rail service from New York or Chicago. Others—like Las Vegas, Reno, Miami, Phoenix, and Tucson—flourished in the air traffic sunshine, so to speak, and attracted more air traffic than their populations suggested because of their status as resorts.[34]

Taaffe's research also demonstrated a strong correlation between the shape of the US urban hierarchy and the air traffic system. Once again he demonstrated that a few cities—New York, Chicago, Los Angeles, and San Francisco—dominated in that they generated the most passenger traffic overall and they "accounted for at least 10 percent of the air passenger traffic" at most other cities as the passengers at the airports in those cities either arrived from one of those four dominant cities or were departing to go to those cities. And their dominance had increased from 1940 to 1954. Further, in addition to these national centers, certain regional centers had also emerged. These included Seattle, Portland, Atlanta, New Orleans,

Dallas, and Houston. And, he argued, the air traffic patterns of the USA seemed to be growing increasingly hierarchical.[35]

Taaffe, however, did not deal with causality, at least not very deeply. He was more interested in creating a model that demonstrated the relationship between air traffic and the urban hierarchy. At base, his was what geographers call a gravity model that showed that larger cities attracted more air traffic. It also offered some explanation as to why some cities had air traffic totals that were either higher or lower than his gravity model would predict—traffic shadow or specialized urban function, for example. Moreover, nothing in Taaffe's work suggested that air transportation had played much of a role in influencing the shape of the urban system. More likely, it generally reflected the existing system.[36] Other studies—both by historians and social scientists—sought to explore more deeply the relationship between airport development and economic development. Major questions emerged such as did airports promote economic development or did economic development create the demand for more and larger airports? And, perhaps equally important, what type of economic development could and should airports attract?

As noted above, one way cities promoted economic development was to seek hub status. Delta Airlines pioneered the hub-and-spoke model after WWII and it used the airport in Atlanta as its primary hub. In an age when direct flights were more the norm, the Atlanta airport emerged as one of the largest transfer hubs in the world as early as the 1950s. Delta also quickly moved into the jet age. Transportation historian Drew Whitelegg argued that the close relationship between Delta Airlines and the city of Atlanta was a significant element in that city's successful economic development activities throughout the post-war period. Delta's corporate commitment to using the Atlanta airport as its hub helped support the city's efforts to expand and improve the airport. The emergence of Atlanta as major transportation center then supported local efforts to promote the city first as a national center and then as an international center for business activity.[37]

A study funded by the Department of Housing and Urban Development (HUD) and conducted by the National Council for Urban Economic Development in 1989 also affirmed the positive economic development role of airports, but was somewhat more qualified. While noting the independent role airports could play in promoting economic development, the study emphasized that cities could maximize the benefits by incorporating their airport expansion plans into a larger, overall economic development strategy. The study especially emphasized the types of real estate developments, such as office and industrial parks that could be attracted to the areas surrounding airports.[38]

Another academic study that explored the relationship between growth in passenger traffic and growth in urban areas complicated the picture by putting it in an historical perspective. Andrew Goetz, a geographer, looked at that relationship before and after the introduction of jet airliners, and before and after the first decade of deregulation. In the early time period—before and after the introduction of jets—he found a strong positive relationship between passenger traffic growth and metropolitan population growth as high-growth cities witnessed higher rates of passenger growth than slow-growth cities. He also found that the relationship involved regional patterns with fast growing Sunbelt cities around the southern and western rim of the nation experiencing high levels of passenger growth while slower growing Rustbelt (eastern and Midwestern) cities experiencing lower levels of passenger growth. Over time and overall, however, the relationship diminished after the 1960s and as one looked at the period of deregulation. Additionally, he questioned claims that airport development independently promoted metropolitan economic development on the grounds that such studies lacked "a conclusive empirical foundation." And he regarded dependence on hubs as risky, because cities that went to the expense and trouble of expanding their airports into a hub might lose both money and air traffic if the airlines serving their airport retrenched or, worse still, failed outright. And, as noted above, events bore him out on this point, for waves of consolidations and mergers in the airlines industry in the 1990s and after resulted in the relocation or closing of many regional and national hubs established during the early heady days of deregulation.[39]

Many airport hubs created in the 1980s not only served as transfer points for domestic flights but also for international flights. These international gateway hubs—including Los Angeles, Miami, and San Francisco—have connections for domestic flights, but tend to focus on the international market.[40] In the late 1980s and into the 1990s, urban sociologist Jerome I. Hodos argued that Philadelphia engaged in a massive airport expansion plan as one part of its overall strategy to maintain that city's position among what Hodos called "second tier" global cities. Working with USAir, which also operated a domestic hub at Philadelphia, the city opened a new international terminal in 1991 and added a second international terminal in 2003. Hodos noted that the airport's fortunes were closely tied to the fortunes of USAir, which twice went into bankruptcy after 2000. Nonetheless, passenger traffic has generally grown and the airport also helped to attract other economic activity to the surrounding area.[41] While it has not witnessed dramatic gains, Philadelphia's airport has ranked consistently as the 19th busiest in 2000, 2005, and 2010.[42]

Hodos emphasized that Philadelphia's airport strategy was just one among many the city pursued to retain its global connections. Others have argued that an airport expansion strategy alone can bring dividends. In a 1991 study, sociologist Michael D. Irwin and professor of business administration and sociology John D. Kasarda emphasized the importance of achieving international gateway status. They challenged the notion that the air transportation network merely reinforced the urban hierarchy created by earlier forms of transportation, such as rail networks and automobile highways and expressways. They argued, instead, that cities developing a central role within the international air transportation network gained "competitive advantages for employment growth." Deregulation, in particular, they argued, created opportunities for upstart metropolitan areas to challenge the centrality of such traditional economic dynamos as New York and Chicago. As airlines established transfer hubs in cities like Denver and Charlotte, those places edged out less fortunate metropolitan areas for access to international markets and the global economy.[43]

From this research, Kasarda coined the term "Aerotopolis" to describe metropolitan areas that experienced the economic boom he associated with international hubs. He claimed that such places appeared first right after deregulation and that more emerged after 2000. Pointing to examples both in the USA and overseas, he detailed examples of rapid economic development and job creation in the vicinity of large, international airports. And he asserted that traffic growth at airports resulted in employment growth rather than employment growth fueling increased traffic growth. Airports were important nodes in the emerging global economy and cities afforded good international connections through their airports had the potential to become that new thing, an Aerotopolis. Kasarda nominated Memphis, Detroit, and Dallas-Fort Worth as the most likely future American claimants to that title. However, he also suggested that cities and airports in the USA were less likely to adopt plans and policies to take advantage the economic development benefits of international airports than were cities overseas. In fact, he said that the type of rapid development that could come with airport expansion could also foster opposition to such development. He pointed to the example of the opposition to the development of the former El Toro airbase in Orange County, California, into a commercial airport.[44] (See Chap. 4, "Response to the Jet Age: Federal–Local Interaction and the Shaping of the Aviation Landscape")

The exact relationship between airport growth and urban growth remains point of debate, though much of the scholarship seems to suggest that airport expansion—absent a larger economic development strategy—was unlikely to bring significant gains. Further, the airport hierarchy has a strong tendency to reflect the urban hierarchy. Regardless, cities have perceived their airports as potential growth machines and have acted accordingly.

Airport Security: It All Began with "Take me to Cuba"

While the hijacking of aircraft was not unknown before WWII, developments in early post-war Europe led to the first high-profile wave of hijackings. Between 1947 and 1953, a number of individuals wishing to escape from behind the Iron Curtain hijacked aircraft as a means to reach sanctuary in the West. Of 16 attempts, 14 succeeded and American officials referred to these individuals as "escapees" not hijackers. However, the imposition of greater travel restrictions in eastern European countries soon closed those routes to the West. A second wave came in 1960 involving individuals from Cuba hijacking airplanes in order to flee to the USA. US officials, again, welcomed the refugees from Castro's regime. At the time, no US officials viewed the possibility of US citizens hijacking planes to fly anywhere as a serious threat. Hijacking was not even a federal crime.[45]

That changed following a series of hijackings in 1961. In May and again in June, US citizens of Cuban origin hijacked US airliners and forced them to fly to Cuba. In the first case, the Castro government immediately seized both the plane and hijacker. It later allowed the plane and passengers to depart after a delay of just a few hours. In the second, Castro again seized the hijacker, but before releasing the plane insisted that the USA return a Cuban naval vessel that had been hijacked to Key West, Florida. These hijacking, plus additional incidents involving disgruntled native-born US citizens, promoted Congress to pass the first anti-hijacking legislation. The Hijacking Act of 1961, an amendment to the Federal Aviation Act of 1958, made the commandeering of an airplane a federal offense and potentially punishable by death. It was hoped that the penalties involved would act as a deterrent.[46]

While hijackings did not stop entirely, they did seem to diminish for a few years. In 1968, however, yet another wave of hijackings imperiled US airlines and their passengers. As was the case with many of the incidents from the early 1960s, a number of the hijackers demanded to be flown to Cuba. This led many officials in the USA to believe that most hijackers were disaffected Cuban refugees wishing to return home. The US State Department even floated the idea of offering such individuals a one-way ticket to Cuba, as long as the former refugee vowed never to return to the USA. Soon, however, it became clear that the hijackers were not all Cuban refugees. Rather, for the most part, they were individuals in trouble with the law wishing to flee to another country. Others were simply mentally unstable. Regardless, most of these hijackings ended peacefully. As a result, both US government officials and airline executives felt the best way to handle the hijackings was not implementation of airport security measures, such as magnetometers to screen for weapons, but to simply give the perpetrators what they wanted, a flight to Cuba. As airlines worked to make such hijackings as painless as possible for passengers, the US public seemed to accept the relatively frequent diversion of US flights to Cuba—more than 20 such incidents in 1968 alone—as routine.[47]

As the epidemic continued into 1969 and the early 1970s, official attitudes toward hijackings changed as two new types of hijackers emerged. First, extremist groups viewed aircraft hijackings as a high-profile way to get their message out to a worldwide audience. Initially, these actions happened outside the USA and involved non-US carriers. Over time, however, US carriers fell victim. For example, in August 1969, members of the Popular Front for the Liberation of Palestine (PFLP) commandeered a TWA aircraft en route from Rome to Tel Aviv. The hijackers forced the plane to land in Damascus, Syria. After freeing the passengers, the PFLP set off dynamite in the cockpit. Among the perpetrators was 25-year old woman, Leila Khaled. She soon became the "face" of the PFLP and something of an international celebrity. In response to these hijackings, the United Nations drafted a multi-lateral anti-hijacking treaty—the Hague Hijacking Convention. But as many nations, including Cuba and countries in the Middle East, proved slow to ratify the treaty, such hijackings continued.[48]

A second new type of hijacker, one seeking to extort money from the airlines, appeared in 1971. One of the most famous was a man known to the American public only as D.B. Cooper. In November 1971, Cooper commandeered a Northwest 727 flying between Portland, Oregon, and

Seattle, Washington. He said that he had a bomb and threatened to blow up the plane if his demands—$200,000 and four parachutes—were not met. When the plane landed in Seattle, authorities gave Cooper his money and the parachutes. Cooper then ordered the pilot to fly to Mexico, with the understanding that the plane would have to refuel en route. Once the pilot made the scheduled landing for fuel in Reno, Nevada, authorities determined that the hijacker had parachuted out the rear stairwell of the aircraft. Despite a massive manhunt, Cooper was never found, though in 1980 a young boy found bundles of $20 bills along the Columbia River with serial numbers matching those of the cash given to Cooper. Cooper became something of a folk hero and the FAA feared there would soon be imitators. And, indeed, by mid-1972, 17 individuals had attempted similar hijackings.[49]

Even before DB Cooper parachuted into legend, the FAA created the first anti-hijacking security measures. The Nixon Administration feared that a lack of success in thwarting hijackings might simply encourage more. Therefore, less than a month after Richard Nixon took office in January 1969, his administration created a Task Force on Deterrence of Air Piracy. The task force focused on three goals: deterring potential hijackers; increasing defenses against hijackings at the nation's airports; and ultimately defeating those who did attempt to hijack aircraft. To deter would-be hijackers, one of the task force's first recommendations was the placement of large signs at airports that outlined the penalties for hijacking and warned passengers that their persons and their carry-on baggage could be subject to search. Airlines, however, balked at the idea, arguing that the signs might just encourage hijackings rather than deter them and at the time neither the FAA nor the airlines had the personnel necessary to search passengers and baggage before boarding. The FAA hoped that the hijackers would not be aware of that fact, but out of deference to the airlines made the posting of the information a voluntary program.[50]

More significantly, the task force also set in motion the creation of the security system that became commonplace at airports throughout the USA for much of the rest of the twentieth century. First, the FAA task force developed a so-called "hijacker profile," consisting of a number of observable behavioral characteristics a potential hijacker might display when attempting to board an aircraft. Second, the task force recommended a series of measures following the identification of an individual fitting the profile. Airline personal would ask any passenger fitting the profile to submit to a pre-board screening involving the use of a magnetometer. If the magnetometer detected a potential weapon, the passenger would then be asked to provide identification. If the passenger refused, a US Marshal would be summoned. A continued refusal would lead to a second screening by the magnetometer. A second positive reading would lead to a physical search of the passenger.[51]

Eastern Airlines, which had experienced a number of hijacking attempts, volunteered to test the new procedures. It did so only at Washington's National Airport, as the primary limitation on the program remained the lack of tested, verified equipment. Throughout 1969, the FAA and Eastern performed tests and gathered data. Though they never ascertained the cause and effect, the number of hijackings decreased during the test period. By the end of 1969, satisfied with the preliminary results, Secretary of Transportation John Volpe announced a modest expansion of the program. Eastern remained the only airline involved and, with equipment still limited, applied the screening procedure only at those airports hijackers seemingly favored and at a few randomly selected lower-threat airports.[52]

Several events in 1970, however, led to the rapid expansion of the modest experimental program. On March 17, 1970, an armed individual hijacked an Eastern Airlines DC-9 flying between Newark and Boston. The hijacker, John J. Divivio, struggled with the flight crew. Shots were fired that killed the co-pilot and wounded Divivio and the pilot, although the pilot managed to land the plan safely. Divivio did not seem to have a destination in mind when he commandeered the aircraft. Since he had previously attempted suicide, some theorized that he had hijacked the airplane and used a weapon as a way to assure his own death. Eastern, not surprisingly, decided to expand the screening program it already had in the light of that incident. The FAA also decided the step up actions.[53]

Meanwhile, the Nixon administration's air piracy task force finished its work. In the spring of 1970, it made a final recommendation that the FAA create a permanent office responsible for aviation security. In response, FAA administrator John Shaffer approved the creation of an Office of Air Transportation Security. Very soon thereafter, September 6, 1970, members of the PFLP, once again including Leila Khaled, hijacked four aircraft (three successfully), including two American airliners, a TWA 707 and a Pan Am 747. The hijackers ordered the Pan Am 747 to fly to Beruit and then to Cairo. Shortly after arrival, and after the passengers and crew were taken off the aircraft, the hijackers blew up the plane. The TWA 707 and a Swissair DC-8 were diverted to an airstrip in Jordan. They were joined by a British Overseas Airway Corporation VC-10, hijacked on September 9. On September 12, 1970, again after releasing the passengers and crew, the PFLP blew up those three aircraft.[54]

Public reaction to that incident was immediate and negative. The Nixon Administration moved quickly to take more extensive measures to prevent hijackings and announced a seven point plan during the fall of 1970. The plan included creation of a federal force of armed officers who would travel on American airliners to thwart would-be hijackers. Military personnel would be used until a group of civilian "sky marshals" could be trained. Federal law enforcement officers would assist in providing the manpower necessary for searches and potential arrests at airports. The federal government would also fund research and development of more effective metal and bomb detecting equipment. The airport screening system already in place, however, remained voluntary. Though more airlines had decided over time to adopt the procedures with which Eastern had been experimenting, the screening system was not yet nation-wide and airlines did not demonstrate a great deal of uniformity in their use and application of the screening procedures.[55]

The White House also announced the creation of a new organization, the Office of Civil Aviation Security, within the DOT. To spearhead the anti-hijacking effort, President Nixon appointed retired Air Force Lt. General Benjamin O. Davis, Jr., to the post.[56] Benjamin O. Davis Jr. had served as one of the Tuskegee Airmen during WWII and came to the job at the DOT from Cleveland, Ohio, where he had a short, but difficult tenure as the director of public safety. Davis oversaw the training of the first class of new air marshals, and he also conducted a tour of airports in the USA and gathered information about security problems.[57]

Davis's research highlighted another weak point in the US anti-hijacking efforts—the physical security of US airports. After a trip to Israel to study security measures at Tel Aviv's airport, Davis concluded that American airport officials should implement programs to keep unauthorized personnel and lay persons out of airport hangars, fuel stations, and boarding areas. Only the federally owned and operated airports—Washington National and Dulles—had airport security programs. Davis recommended the creation of federal regulations to enhance airport security and traveled extensively to promote greater airport security, a campaign that began producing results only in 1973.[58]

In the meantime, air marshals—at first military, and then civilian—began flying on US aircraft. Financed by taxes and fees, the deployment of air marshals was intended to deter hijackers, or failing that, subduing hijackers on aircraft in flight. Though the public may have developed some sense of reassurance from the program, the Nixon administration, particularly FAA administrator Shaffer, hoped the program would be temporary, and ironically, a failure of the program in 1971 contributed to its termination. In October of that year, a hijacker successfully diverted an American Airlines 747 to Cuba, despite the presence on board of three air marshals and a Federal Bureau of Investigation (FBI) agent, all of whom decided that an in-flight armed encounter with the hijacker would be too dangerous to the passenger and crew. Following the implementation of mandatory passenger screening in 1973, the air marshal program was effectively dissolved until its reinstatement in 1985 in response to another wave of aviation directed terrorism.[59]

As early as July 1971, the FAA tried ordering the search of all carry-on baggage on flights that had proved favorite targets for hijackings. Metal detection equipment, though, remained in short supply. Unless airlines conducted a physical search of the passengers and their bags, it was still possible for hijackers to smuggle weapons on board aircraft as only passengers identified through the hijacker profile had been subject to physical search. It seemed, however, that requiring all passengers to pass through a metal detector before boarding might be the only way to prevent hijackings. That type of indiscriminate search was potentially unconstitutional. The FAA, nonetheless, ordered 1500 metal detectors in anticipation of eventually moving toward the universal screening of passengers and their carry-on luggage.[60]

In 1972, an election year, a series of high profile hijackings put great pressure on Congress and the Nixon Administration to take additional action. These included the July 5 hijacking of a Pacific Southwest Airlines jet that resulted in the first death of a US passenger during a hijacking when the FBI decided to storm the plane once it landed in San Francisco and a November 19 hijacking of a Southern Airlines flight during which one of the perpetrators threatened to "bomb" the nuclear reactor at the Oak Ridge National Laboratory.[61] In December 1972, FAA Administrator Shaffer issued an emergency order that required the screening of all passengers and carry-on baggage—either mechanically or by hand—effective January 3, 1973. Further, the FAA required all airport operators to submit a security plan and all plans had to include the stationing of armed guards at all boarding check points by February 1973.[62]

The implementation of that program immediately raised two issues. First, were such searches of all passengers constitutional, a question that was answered in the affirmative in a judicial decision rendered in June 1973.[63] The other issue involved who would provide and pay for the added security personnel. The airlines and local airport operators argued that preventing hijackings was a federal responsibility. They favored the expansion of the air marshal program or the creation of a new federal law enforcement program to provide the needed guards. The Nixon Administration, on the other hand, viewed the passenger screenings as a local responsibility and insisted on paying for it with an airport-user (passengers and airlines) funded program. The Senate sided with the airlines and local authorities when it passed a bill that included the creation of a federal airport security guard program in February 1973. The House, however, which had failed to act on a similar bill the year before, did not pass an anti-hijacking bill of its own in 1973. After a violent hijacking attempt at Baltimore's airport—which included a scheme by the hijacker to crash the plane into the White House—in March 1974, the House finally passed a bill, but it did not include the federal airport security guard program. A year's worth of experience with a system that was locally funded and operated, however, helped convinced Senators to drop the provision calling for a federal security force. President Nixon signed the Antihijacking Act of 1974 into law on August 5, 1974, three days before he resigned the presidency.[64]

The basic security framework that would become so familiar to American air travelers prior to September 2001 was thus in place by the early 1970s. Metal detectors, X-ray machines (for carry-on bags), and armed guards all became ubiquitous at America's airports. At the time, it ranked as one of the most rigorous aviation security systems in the world. Additional small changes would come in the wake of renewed hijackings in the 1980s and 1990s, but the basic system was largely unaltered. And the security system remained the responsibility of local airport officials working with the airlines. Only the dramatic and traumatic events of September 11, 2001, would prompt a significant change in both the framework and its funding structure.

Conclusion

Jet airliners, deregulation, rapid growth in the number of Americans flying, economic development hopes, and security fears all shaped US airports in the post-war period. Although how each of these played out varied from place to place, generally speaking between 1945 and the early twenty-first century, America's commercial airports became larger and more complex. In addition, two other important factors are also important to understanding the history of US airports since 1945. The first is that airports in the USA are locally owned and managed. And although there have been often high-profile calls for privatization, for the most part the commercial airports in the US remain local, public property. Second, airports have received much of their funding for operations and improvements from local and private (airline) sources. However, as parts of a national and even international system of transportation, a key to unlocking both those sources of funding has been the monies provided by the federal government. Though frequently a focus of intense debate, federal funds have generally flowed to the nation's airport, especially since the 1970s and the creation of the Airport Trust Fund. The ownership and funding structure is the subject of the next chapter.

Notes

1. John D. Anderson, *The Airplane: A History of Technology* (Reston, VA: American Institute of Aeronautics and Astronautics, 2002), 285–290.
2. Roger E. Bilstein, *The American Aerospace Industry: From Workshop to Global Enterprise* (New York and London: Twayne Publishers, an imprint ofSimon & Schuster McMillan and Prentice Hall International, 1996), 141.
3. Anderson, *The Airplane*, 349–351.
4. For a discussion of the Constellation, see http://aerostories.free.fr/connie/page10.html (accessed August 5, 2014). For a discussion of the 707, see http://history.nasa.gov/SP-468/ch13-3.htm (accessed August 5, 2014).
5. Erik M. Conway, *High Speed-Dreams: NASA and the Technopolitics of Supersonic Transportation, 1945–1999* (Baltimore: The Johns Hopkins University Press, 2005), 132–133.
6. Anderson, *The Airplane*, 338–341.
7. Dilip. R. Ballal and Joseph Selina, "Progress in Aero Engine Technology (1939–2003)" in "Centennial of Powered Flight Celebrations," Special Issue on Air Transportation, AIAA Journal of Aircraft, 2003, 7, 17.

8. Bilstein, *The American Aerospace Industry*, 177.
9. Bilstein, *The American Aerospace Industry*, 178; James Hansen, *The Bird is on the Wing, Aerodynamics and the Progress of the American Airplane* (College Station: Texas A&M University Press, 2004), 187.
10. Hansen, *The Bird is on the Wing*, 188–189; David T. Courtwright, *Sky as Frontier: Adventure, Aviation and Empire* (College Station: Texas A&M University Press, 2005), 145.
11. For information on the 707, see http://www.boeing.com/boeing/commercial/707family/index.page (accessed August 5, 2014). For information on the 747 see http://www.boeing.com/boeing/commercial/747family/specs.page? (accessed August 5, 2014).
12. For a concise version of the history of the US SST, see Hansen, 143–174. For a more complete examination of the program to create a US SST, see Conway, *High Speed Dreams*, 118–188.
13. Mark F. Rose, Bruce E. Seely, Paul F. Barrett, *The Best Transportation System in the World: Railroads, Trucks, Airlines, and American Public Policy in the Twentieth Century* (Columbus: The Ohio State University Press, 2006), 76–83; Paul Stephen Dempsey and Andrew R. Goetz, *Airline Deregulation and Laissez-Faire Mythology* (Westport, CT: Quorum Books, 1992), 159–171; Richard H. K. Vietor, "Contrived Competition: Airline Regulation and Deregulation, 1925–1988" *The Business History Review* 64 (Spring 1990): 68–74.
14. Vietor, "Contrived Competition," 68–74.
15. Rose, Seely, Barrett, *The Best Transportation System in the World*, 83–96, 186–187; Vietor, 74–83.
16. Rose, Seely, Barrett, 151–211; Phillip J. Cooper, *The War Against Regulation: From Jimmy Carter to George W. Bush* (Lawrence: University of Kansas Press, 2009), 14–28.
17. Vietor, 82–83; Rose, Seely, Barrett, 187–188. Dempsy and Goetz, *Airline Deregulation and Laissez-Fair Mythology*, 179–191.
18. For the complete story of the PATCO strike, see Joseph A. McCartin, *Collision Course: Ronald Reagan, the Air Traffic Controllers, and the Strike that Changed America* (New York: Oxford University Press, 2011); Vietor, p. 98.
19. Paul Stephen Dempsy and Andrew R. Goetz, 199–203.
20. Dempsy and Goetz, 199–209.
21. Vietor, 96–99.
22. Vietor, 99–100.

23. Andrew R. Goetz and Christopher J. Sutton, "The Geography of Deregulation in the U.S. Airline Industry" *Annals of the Association of American Geographers* 87 (June 1997): 243.
24. Goetz and Sutton, "The Geography of Deregulation," 245–246.
25. Goetz and Sutton, 242–260.
26. Examples of studies of cities and transportation technologies include Ann Durkin Keating, *Chicagoland: Cities and Suburbs in the Railroad Age* (Chicago: University of Chicago Press, 2005); Mark Foster, *From Streetcar to Superhighway: American City Planners and Urban Transportation* (Philadelphia: Temple University Press, 1981); Charles N.Glaab, *Kansas City and the Railroads: Community Policy in the Growth of a Regional Metropolis* (Lawrence: University of Kansas Press, 1993); Peter J. Ling, *America and the Automobile: Technology, Reform and Social Change* (Manchester: Manchester University Press, 1990); McShane, *Down the Asphalt Path*; Sam Bass Warner, Jr., *Streetcar Suburbs: The Process of Growth in Boston (1870–1900)* (Boston: Harvard University Press, 1962); Martin Wachs and Margaret Crawford, eds., *The Car and the City: The Automobile, the Built Environment and Daily Urban Life* (Ann Arbor: University of Michigan Press, 1992).
27. "Annual Results US Airlines" http://airlines.org/data/annual-results-u-s-airlines-2/ (accessed June 20, 2016).
28. For top ten 1954, see "Chicago's Midway Nation's Busiest Field," *Aviation Week* 62 (April 11, 1955): 132; for top ten 1966, see "The Top Two Dozen—1966" *Flight International* (October 5, 1967), 566; for top ten 1976, see Drake Edward Warren, "The Regional Economic Effect of Commercial Passenger Service at Small Airports," Ph.D. diss., University of Illinois at Urbana-Champaign, 2008, p. 32.
29. For the top ten airports in 1985, see Federal Aviation Administration, Department of Transportation, "U.S. Airport Enplanement Activity Summary for CY 1985 Listed by Rank Order, Enplanements," 30 September 1986;
30. For the top ten airports in 1985, see Federal Aviation Administration, Department of Transportation, "U.S. Airport Enplanement Activity Summary for CY 1985 Listed by Rank Order, Enplanements," 30 September 1986; for top ten airports 1995, see Federal Aviation Administration, Department of Transportation, "Changes in Revenue Passenger Enplanements at Primary Airports Base Year CY 1984 (By Rank Order CY 1985)," October 16, 1996; for top ten airports 2005, see Federal Aviation Administration/Department of Transportation,

"Calendar Year 2005 Primary and Nom-Primary Commercial Service Airports," October 31, 2006; for top ten airports 2010, see Federal Aviation Administration/Department of Transportation, "Enplanements at Primary Airports (Rank Order) CY 10," October 25, 2011.

31. Department of Transportation, "Enplanements at Primary Airports (Rank Order) CY 10," October 26, 2011.
32. Thomas J. Sheeran, "Cleveland Hub Closing," *USA Today*, February 4, 2014; Stacy Vogel Davis, "Re-routed: Diminished airline service forces Milwaukee businesses to seek alternatives," *Milwaukee Business Journal*, December 21, 2012; Sarah Hale, "Kansas City is Doing Just Fine without a Major Airline Hub," *Wall Street Journal*, July 2, 1999.
33. For a recent work examining the gap between promises and the actual economic development performance of a popular local project, the convention center, see Sanders, *Convention Center Follies.*
34. See Edward J. Taaffe, "Air Transportation and United States Urban Distribution," *Geographic Review* 46 (April 1956): 219–238.
35. Edward J. Taaffe, "The Urban Hierarchy: An Air Passenger Definition," *Economic Geography* 38 (Jan., 1962): 1–14.
36. Andrew R. Goetz, "Air Passenger Transportation and Growth in the U.S. Urban System, 1950–1987," *Growth and Change* 23 (Spring 1992): 217–220.
37. See Drew Whitelegg, "Keeping their eyes on the skies: Jet aviation, Delta Air Lines and the growth of Atlanta," *Journal of Transportation History* 21 (March 2000): 73–91.
38. Sarah Eilers, "Airport Growth: Creating New Economic Development Opportunities," (Washington, D.C.: National Council for Urban Economic Development, December 1989).
39. Ibid., 235–236. For a more recent study of the relationship between airports and economic growth, see Carine Discazeaux and Mario Polese, "Cities as Air Transport Centres: An Analysis of the Determinants of Air Traffic Volume for North American Urban Areas," (Montreal: Institut national de la rechereche scientificque Urbanisation, Culture et Societe, working paper, November 2007).
40. Goetz and Sutton, 243.
41. See Jerome I. Hodos, *Second Cities: Globalization and Local Politics in Manchester and Philadelphia* (Philadelphia: Temple University Press, 2011), esp. 140–142.
42. See Federal Aviation Administration, Department of Transportation, "Primary Airport Enplanement Activity Summary for CY 2000 Listed

by Rank Order, Enplanements," October 19, 2001; Federal Aviation Administration/Department of Transportation, "Calendar Year 2005 Primary and Non-Primary Commercial Service Airports," October 31, 2006; Federal Aviation Administration/Department of Transportation, "Enplanements at Primary Airports (Rank Order) CY 10," October 25, 2011.

43. See Michael D. Irwin and John D. Kasarda, "Air Passenger Linkages and Employment Growth in U.S. Metropolitan Areas," *American Sociological Review* 56 (August 1991): 524–537.
44. See "Planning the 'aerotropolis'," *Airport World* 5 (Oct-Nov 2000): 52–3; "Airport Cities Drive Local Economic Development," *MuniNetGuide* (March 23, 2010). http://www.muninetguide.com/print.php?id=361 (accessed July 13, 2010). For a critical analysis of the "aerotropolis" model for urban development, see Robert Freestone, "Planning, Sustainability and Airport-led Urban Development," *International Planning Studies* 14 (May 2009): 161–176. For another discussion of airports and economic development see Robert Bruegmann, "Airport City," in John Zukowski, ed., *Building for Air Travel: Architecture and Design for Commercial Aviation* (Munich and New York: Prestel-Verlag; Chicago: The Art Institute of Chicago, 1996), 195–211. And for a longer discussion of Kasarda and his ideas about future airports and the "aerotropolis" see Kasarda and Lindsay, *Aerotropolis.*
45. Robert M. Hardaway, *Airport Law, Regulation and Policy* (New York: Quorum Books, 1991), 136; Brendan I. Koerner, *The Skies Belong to Us: Love and Terror in the Golden Age of Hijacking* (New York: Random House, 2013), 33–37.
46. Hardaway, *Airport Law*, 136; Koerner, *The Sky Belongs to Us*, 37–38.
47. Richard J. Kent, *Safe Separated and Soaring: A History of Federal Civil Aviation Policy, 1961–1972* (Washington, D.C.: U.S. Department of Transportation, Federal Aviation Administration, 1980), 333: Douglas M. Kraus, "Searching for Hijackers: Constitutionality, Costs, and Alternatives," *The University of Chicago Law Review* 40 (Winter 1973): 385; Koerner, 41–50.
48. Koerner, 50–52.
49. Kent, *Safe, Separated and Soaring*, 346–47.
50. Ibid., 338–9.
51. Kent, 339; Hardaway, 136–137.
52. Kent, 339–342.

53. Ibid., 342–343.
54. Koerner, 74–75; Alastair Gordon, *Naked Airport: A Cultural History of the World's Most Revolutionary Structure* (New York: Metropolitan Books, Henry Holt and Company, LLC, 2004), 231–232.
55. Kent, 343–344; Kraus, "Searching for Hijackers," 388; Alona E. Evans, "Aircraft Hijackings: What is Being Done?" *The American Journal of International Law* 67 (October 1973): 648–649.
56. Kent, 343–344.
57. Benjamin O. Davis, Jr., *Benjamin O. Davis, Jr.: American, An Autobiography* (Washington, D.C.: Smithsonian Institution Press, 1991), 352–357.
58. Davis, *Benjamin O. Davis, Jr.*, 358–361; Evans, "Aircraft Hijackings," 369; "Airport Security Measures Studied," *Aviation Week & Space Technology* 93 (November 9, 1970): 28.
59. Kent, 345–346.
60. Ibid., 348–349; "Airline, Airport Security Drive Still Faces Financial Obstacles," *Aviation Week & Space Technology* 96 (March 20, 1972): 28–29.
61. Koerner, 185–187, 203–207.
62. Kent, 350–351; Edmund Preston, *Troubled Passage: The Federal Aviation Administration During the Nixon-Ford Term 1973–1977* (Washington, D.C.: Department of Transportation, Federal Aviation Administration, 1987), 38–39.
63. Preston, *Troubled Passage*, 42.
64. Preston, 41, 52–55; "Airports Seek to Shift Security Burden," *Aviation Week & Space Technology* 98 (January 22, 1973): 23.

Chapter Two: Closer to Ground Level: Airport Ownership and Finance

US airports form vital—indeed indispensable—parts of a national and even international system of air transportation. However, beginning in the earliest days of commercial aviation in the late 1920s, local governments increasingly owned and managed them. And since at least the 1930s, those local governments, recognizing the role their airports played in the emerging national air transportation system, called for help from the federal government. The federal government, on the other hand, has generally emphasized the local nature of the nation's airports. Many federal officials have insisted that much of the funding for airport expansion and improvement come from local sources as well as from the private corporations—the airlines—that use those facilities. Nonetheless, enough federal officials recognized the national/international role of airport to create a system of federal funding based on user taxes, similar to the federal system of funding for the nation's highways. Understanding this combination of local ownership and federal funding is important to understanding the post-war development of US airports.

Who Owns the Airport?

During aviation's formative years, ownership of airports was quite varied. From World War I (WWI) into the late 1920s and early 1930s, many so-called "municipal airports" were under private ownership and management. Chambers of Commerce and other local business organizations as

J.R. Bednarek, *Airports, Cities, and the Jet Age*,
DOI 10.1007/978-3-319-31195-1_3

well as specially created private corporations often acted to establish local airports either in the absence of needed state enabling legislation or in the light of municipal inaction, perceived or real. As commercial aviation grew and airports became more complex—and hence more expensive—private ownership became less attractive. The key to the shift from private to public ownership, however, came with the New Deal. Programs such as the Civil Works Administration (CWA), the Public Works Administration (PWA), and the Works Projects Administration (WPA), all of which provided aid for airport improvements, required a certain level of public involvement before a facility could be eligible for federal monies. For example, the earliest program, the CWA, allowed for private ownership of airports receiving aid, but required that the airport be leased by a public entity such as a local government. The later WPA, however, required public ownership of any airport receiving federal aid.[1]

In the early twenty-first century, local governments, primarily cities or counties, remained the most common owners and managers of the nation's air carrier airports. Of the nearly 550 public airports certified for commercial airline traffic in the USA, cities or towns own or manage approximately 180, with multiple cities or city/county combinations owning or managing jointly an additional handful. Counties alone own and manage about 100. Other less common forms of public ownership and management include that by states, as both Alaska and Hawaii have extensive state-owned airport systems, with all the air carrier airports in the latter under state ownership. The states of Arizona, Maryland, Maine, Montana, New York, North Carolina, Rhode Island, and Vermont each own one airport, while Connecticut owns two. There are also a number of joint civilian/military facilities, while public universities (University of Illinois, Purdue University, University of Mississippi, Ohio State University, Penn State University, and Texas A&M University) own and manage six airports.[2]

The second most common form of airport ownership and management, still essentially local in nature, is that by an independent public authority.[3] Proponents of this type of ownership generally argued public authorities could bring more focused and professional management to the airports. Cities that transferred ownership of their airports to public authorities generally did so due to local resistance to funding airports with general obligation bonds issued against a city's bonded indebtedness limit. Independent authorities could not only issue revenue bonds, they most often could do so without a local vote. Further, in several cases, supporters of airport development often, though certainly not always, called for a regional approach to the issue. In this, they echoed the efforts of many across urban American seeking to find a framework that would allow them to affect a measure of centralization, an "upward shift in the locus of authority for shaping the urban-political economy."[4] Not only did local efforts to create regional airport authorities reflect the push toward regional planning, but the initial role of the federal government in several instances demonstrated the extent of centralization present by the 1960s. For example, in the early 1960s, both the FAA and the CAB came out in favor of regional airports and regional airport planning.

However, as was the case in most efforts in the 1960s and 1970s, champions of a shift to authority ownership succeeded more often in simply changing the ownership or management structure. They were not successful in promoting regional airport planning, reflecting similar failures elsewhere. Though supported by certain political and economic elites, efforts to create regional airport authorities came under sharp criticism from other political and economic elites, especially those far more focused on local concerns.[5] And at the federal level, especially after 1972, the Nixon and Ford administrations developed policies "to replace the politics of centralization with the politics of devolution."[6] As result, by the early 1970s, the federal government largely decided to leave decisions on airport ownership, management, and planning at the local level.

Though a number of airport authorities around the country dated to the 1940s, the two prominent early examples of the transfer of airports from cities to independent public authorities involved, first, a specialized airport authority (or commission as it was called under Minnesota law) and, second, an existing public authority, the Port Authority of New York and New Jersey (PNYNJ). In the first case, the twin cities of Minneapolis and St. Paul each had a commercial airport established in the 1920s. The city of St. Paul owned and managed the airport there, while in Minneapolis, the Board of Park Commissioners owned and managed the airport.[7]. During the 1920s and 1930s, both cities engaged in heated competition for commercial air traffic. In 1943, the state legislature stepped in to stop the battle, creating the Minneapolis-St. Paul Airport Commission, charged with assuming "jurisdiction over all airports within 25 miles of the city hall of either participating city." The ownership of the existing airport properties was turned over to the new commission "without compensation," an action later upheld by the Minnesota Supreme Court. That decision further asserted that to the extent possible the state could and should exert “a unified, integrated, centralized system of control of all classes of aerial traffic in the common ‘air ocean’ over the state” as the creation of “a governmental corporate instrumentality to own and control all airports in a metropolitan area” was legal and justified.[8]

About the same time the state of Minnesota created the organization to own and manage the airports in the Minneapolis-St. Paul area, the PNYNJ seriously began to consider taking over the operation of the two existing air carrier airports in the New York area—Newark and La Guardia—as well as a third airport, then just under construction, Idlewild (later JFK). Between 1943 and 1947, the Port Authority carefully laid the groundwork necessary to gain control of the region’s air carrier airports. The Port Authority faced often strong opposition from some elected city officials as well as from long-time opponents to the authority itself. Some the strongest opposition came from Robert Moses, the controversial “master builder” and himself the head of numerous public authorities, in New York City. Determined to thwart the Port Authority, Moses persuaded the New York State legislature to create a competing New York City Airport Authority to take over ownership and management of the air fields within the city. In response, the Port Authority emphasized the consequences of past municipal airport competition[9] and argued that the new authority could not acquire the financial backing to further develop LaGuardia and, moreover, to adequately develop Idlewild. The Port Authority also asserted that it was the most likely organization to provide the professional, comprehensive, and coordinated planning needed to fully and efficiently provide for the post-war aviation needs of the region. In addition, the Port Authority built support with some local government officials, the Regional Planning Association, the airlines, the federal government, and much of the local press. In the end, a combination of factors, including the vast financial resources of the Port Authority, swung the argument in favor of its assumption of responsibility for the metropolitan area’s major airports.[10] The PNYNJ holds long-term leases on the Newark Liberty, LaGuardia, JFK airports—three of the busiest in the nation—and, since late 2007, Stewart International Airport. The port authority also owns Teterboro Airport, a general aviation airport.

In the early twenty-first century, independent authorities own and/or manage 15 of the 50 busiest airports in the USA and overall entities explicitly named as either airport or port authorities own or manage approximately 128 of the nearly 550 public, commercial airports, with approximately two dozen additional under various related entities called independent boards, commissions, or districts.[11] All but 12 states have one or more airports under authority ownership and/or management. The largest numbers of airport authorities controlling such facilities are in Florida (12), Pennsylvania (11), Illinois (9), Indiana (8), California (7), Montana (7), and West Virginia (7).[12] States with no airport authorities include Alaska and Hawaii (both with high levels of state ownership), the small states of Connecticut and Rhode Island, as well as Idaho, Maryland, Maine, Missouri, New Mexico, Utah, Vermont, and Wisconsin.[13] In Texas, the Dallas-Fort Worth International Airport is managed by a semi-autonomous board, but jointly owned by both Dallas and Fort Worth, as both cities refused to surrender ownership to an independent authority.[14] Though the FAA lists the Metropolitan Washington Airport Authority as the owner of Reagan Washington National and Dulles International, it actually leases both airports. Ownership remains with the federal government, which originally built and developed both airports.[15] They remain the only airports not locally owned, though now locally managed, in the USA.

Privatization: From Thatcherism to September 11, the Road Lightly if at All Traveled

Another alternative form of airport ownership, privatization, emerged in the late 1980s and early 1990s. It reflected a number of factors including long-standing efforts to transfer ownership of the Washington area airports from the federal government as well as the Reagan Administration's embrace of market-based solutions and its campaign to shrink the size of the federal government through privatization of functions. Also involved were the privatization initiatives of the British government under Prime Minister Margaret Thatcher. Although many insisted upon full privatization, the federal government essentially stopped short of that, agreeing to experiment with the private management of commercial airports while making full privatization very difficult. Also, the nation's airlines, which paid the landing fees and gate rentals that essentially provided the bulk of airport revenue, rejected privatization, fearing that it would increase their costs. Thus, despite many strong proponents and increased examples of privatization internationally, in the US privatization remained minimal at the beginning of the twenty-first century.[16]

In 1966, the British government created the British Airport Authority (BAA) to own and manage four commercial airports (Heathrow, Gatwick,

Stansted, and Prestwick) previously controlled and operated by the Ministry of Defence. Over time the BAA also gained responsibility for three Scottish regional airports (Glasgow, Edinburgh, and Aberdeen). From the beginning, the BAA proved a profitable operation. In the mid-1980s, the Thatcher government explored the possibility of privatizing the BAA. First, the privatization would bring a windfall to the British government. Second, a government study suggested that the more flexible and innovative management possible under privatization might prove successful in providing additional capacity at Britain's commercial airports, a very controversial issue. Following the favorable recommendations made in a government white paper, the British government offered BAA for sale in 1986. Although the authority and its airports were privatized, the British government retained certain regulatory controls to address issues surrounding BAA's monopoly over commercial air service in Britain.[17]

At the same time the British government was moving toward the full privatization of its major commercial airports, the idea of privatizing Washington National and Dulles arose in response to proposed legislation to turn those airports over to a regional airport authority. The legislative effort came following the recommendation of an advisory committee appointed by Secretary of Transportation Elizabeth Dole, which included many of the airport users who had opposed any change in ownership in the past, including the Air Transport Association (ATA), the industry group representing the nation's airlines. The legislation initially proposed a 35-year lease (later raised to 50 years) between the government and the new regional airport authority. The lease would cover not only the airport property, but also the access highways and any other related facilities. All airport employees would transfer to the new authority. The legislation proposed that over the lease period, the authority pay the federal government $47 million "for capital improvements made over the years and not reimbursed to the Treasury from airport revenues." Further, the new authority "would assume the estimated $35 million in unfunded future liability for retirement costs of personnel employed at the time of the transfer."[18]

While the airlines supported the authority legislation, opposition came from Senator Gordon J. Humphrey (R-New Hampshire) and the National Taxpayers Union. Senator Humphrey proposed that instead of leasing the airports to a regional authority for $47 million over 50 years, the government seriously explore the possibility of selling the airports to a private buyer. He estimated that such a sale could bring in as much as $1 billion. In fact, a group of private investors, the Merchant Bank of N. M. Rothschild & Sons, Ltd., (NMR), was interested in exploring the idea. NMR argued that placing the airports under private ownership would result in needed upgrades, greater financial viability, and greater contributions to the regional economy. The other arguments favoring privatization came from the National Taxpayers Union, which also focused on the lease terms. The organization felt that at $47 million over 50 years, the government was essentially giving away the airports. Both Humphrey and the National Taxpayers Union wanted Congress to amend the legislation to allow for the sale of the airport to the highest bidder.[19] This opposition, though, came very late in the legislative process and, despite some Reagan Administration support for full privatization, the legislation passed creating the Washington Metropolitan Airports Authority. Opponents had managed to change some of the lease terms, however. Under the final legislation, the new authority agreed to pay $150 million (in 1986 dollars) over the 50-year life of the lease.[20]

Even though the call to privatize National and Dulles failed, the idea of privatizing commercial airports in the USA had a significant national airing as a result. Within a few years, several new proposals to privatize commercial airports arose, once again putting the issue on the agenda. The incident that eventually set in motion the creation of a federal policy to allow some limited privatization of commercial airports in the USA came in 1989 when Albany County, New York, asked the FAA for permission to lease its airport in the wake of two proposals it had received.[21]

Albany County had proposed leasing its airport to either another public entity or a private company as a way to gain needed revenues for the county. At the time, federal regulations dealing with airport grants all but precluded the outright sale of the airport. They also prohibited the diversion of airport revenues to non-airport purposes. County officials saw the lease option as a way to bring airport revenues, in the form of lease payments, into county coffers. As it was, the airport generated a profit of about $4 million per year, all of which under federal rules had to be used by and for the airport. The FAA rejected the lease proposals because the offer from the Lockheed corporation—which had owned and managed the Burbank airport (now Bob Hope) until its sale to a new airport authority in 1978 and managed a number of small airports in the USA—did not guarantee that revenues generated would be used to finance improvements at the airport. There were also concerns over whether Lockheed would operate the airport in a way that best served the public. The other lease proposal came from the local transit authority. According to the FAA, the transit authority did not demonstrate how it would generate the revenues needed to make the lease payments. In both cases, the FAA's primary concern was to guarantee that airport revenues would be used to maintain and improve the airport.[22]

The Albany proposal, though initially rejected,[23] prompted Secretary of Transportation Samuel Skinner to create a task force to examine the issue of privatization. After studying the issue for at least year, in March 1990, the task force reported that it could not provide the secretary with recommendations on establishing a policy on privatization. That task force and its working groups instead identified a number of issues that needed to be addressed before privatization could proceed. Those issues included "ensuring the proper use of federal airport funds, guaranteeing compatible use of land on and adjacent to the airports and preventing excessive profits by the operators." Moreover, answers to the questions involving how privatization might benefit the public, the airlines, and general aviation seemingly eluded the task force.[24]

Despite the failure of the task force to provide policy recommendations, the idea of privatizing the nation's commercial airports gained additional attention when in May 1990 Ronald S. Lauder, heir to the cosmetics fortune and a former US ambassador to Austria who had unsuccessfully sought the mayor's office in New York in 1989, published an opinion piece supporting the privatization of New York City's airports in the *New York Times*. Emphasizing the funds a sale would bring to the fiscally challenged city, Lauder argued that under the PNYNJ the airports failed to generate the revenues they could if they were operated as a business. He rejected the idea that the city take over the airports from the port authority stating that the city was no more capable of operating the airports than the port authority. Instead, Lauder emphasized the sale to a private entity, citing the example of the BAA. Such a sale, he argued, would bring an immediate windfall to the city. Further, in private hands, the airports would once again be on the city's property tax rolls. Property taxes could generate as much as $212 million per year in city revenues. The state and the federal government would also benefit from tax payments. He strongly encouraged the final development of a policy to fully privatize commercial airports in the USA.[25]

By the time of the Albany proposal and the Lauder op-ed piece, a number of academics and public policy analysts had or soon would advocate the privatization of commercial airports in the USA. In 1987, for example, Gabriel Roth, a transportation policy analyst, published a paper on airport privatization in *Proceedings of the Academy of Political Science*. The paper examined the debate surrounding the fate of the Washington area airports as well as offering a brief examination of proposals for privatization. According to Roth, the Heritage Foundation called for the sale of National and Dulles as early as 1984 and Robert W. Poole, Jr. of the Reason Foundation then repeated that recommendation in 1985. Poole had calculated the capital value of the airports at $2.5 billion. Roth expressed a certain level of disbelief that the Reagan Administration, which had "committed to economic efficiency in the use of public resources, to reduction of Federal deficits, and to the encouragement of private enterprise," rejected the idea of turning the airports over to the private sector. Though offering no exact reason why the Reagan Administration decided against privatization, the author suggested that both the FAA and Congress simply failed to grasp the economic benefits of privatization to the government.[26]

For whatever reasons, despite the apparent success of the British model and calls for privatization in a number of quarters, the Reagan Administration did not act to privatize airports. The privatization idea, however, continued to have support from economists, policy analysts, and some local officials desperate to enhance local revenues. Under continued pressure to at least explore the idea, on April 30, 1992, President George H.W. Bush issued Executive Order 12803, which created the rules under which local and state governments could privatize public assets that had received federal dollars. This was followed by the Clinton Administration Executive Order 12893, which aimed at lowering federal barriers to privatization at the state and local levels. A Government Accounting Office (GAO) study determined that little action to privatize any public assets came as a result of those initiatives. In 1996, however, Congress created a more specifically focused airport privatization pilot program as part of the Federal Aviation Reauthorization Act. The new program, though, was a far cry from the full privatization called for by many advocates. Instead, while not ruling out full privatization of commercial airports at some time in the future, it allowed for only experimentation with public–private partnerships at a limited number (five) of airports, both commercial and general aviation. The commercial airports in the program would remain under public ownership, but management could be contracted to a private firm, reflecting the reality that many airport services were already contracted out to private firms. Airline companies continued to oppose full privatization, fearing that private management firms would increase fees in order to maximize profits.[27]

The pilot program came in the wake of a public debate over privatization of the management of the airports serving New York City. In 1994, the new mayor, Rudolph Giuliani, focused on improving the fiscal health of the city, proposed ending the port authority's lease on the city's two airports and turning management over to a private company. At the time, Giuliani was negotiating changes in the lease structure with the port authority. The existing leases were not due to expire until 2015, but the issue came to a head when the port authority, engaged in extensive renovations at the airports in the 1990s, cut the annual lease payments from $600 million per year to the minimum allowed under the lease terms, $3.1 million. Giuliani's initiative also had the support of New York's Republican governor, George Pataki. The governor agreed that the port authority payments to the city were inadequate and, in fact, he favored full privatization of the airports. The proposal generated a great deal of press in late 1994 and early 1995, as, for example, the *New York Times* printed an editorial that opposed private management of the airport and expressed doubts on the sale of the airport land to private investors. Despite the high profile in the press, no immediate action ensued.[28]

The idea of privatizing the management of the airport, however, remained alive. And it gained viability following the creation of the federal privatization pilot program. In 2000, Mayor Giuliani renewed his fight against the Port Authority by accepting bids from four companies for consulting services as the city continued to explore its options. The city contracted with the BAA's US subsidiary (BAA U.S.A.) to help it develop plans for an eventual takeover of the airports.[29] In the end, though, the Port Authority maintained its control of the airports. Giuliani's successor, Michael Bloomberg, briefly explored another option—a land-swap deal in which the port authority would get the land under the airports and the city the World Trade Center site—but in 2003 the city and the port authority agreed to a lease extension to 2050.[30]

After a great deal of fanfare and expectations of widespread action, only a few limited examples of privatization existed in the USA early in the twenty-first century, including the airports around Indianapolis. The Indianapolis Airport Authority owns five airports—one commercial and four general aviation airports. In 1995, it contracted with BAA U.S.A. to manage its system for a 10-year period. The early results were positive, and in 1998, BAA U.S.A. reported that it had exceeded its financial goals. The Indianapolis Airport, however, remained the largest airport authority to privatize its management, and although the management contract was initially extended to the end of 2008, in 2007 the airport authority decided to terminate the contract early.[31] The parting was described as mutual. The year before a Spanish construction firm, Ferrovial, had purchased BAA U.S.A.'s parent company, BAA, and announced that it would concentrate on BAA's airport assets in the UK.[32]

In addition to Indianapolis, the Orlando Stanford International Airport operates under a public/private partnership of the Sanford Airport Authority and Airports Worldwide, a multinational corporation. Airports Worldwide owns TBI Airports Management (TBI AM), originally a British company which, beginning in 1997, managed the international and domestic terminals, recruited additional air carriers, and oversaw ground handling and cargo services at Orlando. TBI AM also has less extensive management consultant contracts with the local governments or authorities responsible for Hartsfield-Jackson International Airport, Bob Hope Airport (Burbank), the Middle Regional Georgia Airport and Herbert Smart Downtown Airport (Macon, Georgia), and Raleigh Durham Airport.[33]

Thus, further complicating the privatization debate in the USA is the fact that many of the companies involved are foreign owned, with sometimes shifting ownership. For example, in 2005, Albertis Airports, a Colombia-based company, acquired British-based TBI AM, which it then sold to Airports Worldwide in 2013.[34] And in 2006, as noted, BAA, whose subsidiary BAA U.S.A. had taken over management of the Indianapolis airport, was sold to Ferrovial, a Spanish construction firm, which outbid Goldman Sachs, the US investment bank.[35] However, the involvement of foreign companies in the management of US airports never became as hotly controversial as did the involvement, post-September 11, of foreign firms in the management of US seaport facilities. In 2006, DP World, a company owned by the ruling family in Dubai, sought to purchase a British company, P&O Ports, that held the contract to operate ports in New York, New Jersey, Baltimore, New Orleans, Miami, and Philadelphia. A political firestorm in the USA forced the company to back out of the purchase deal. Nonetheless, at the same time foreign firms retained management of other US port facilities, including 80 % of the terminal at the Port of Los Angeles.[36] And foreign firms continued to hold management contracts at several US airports, although some of those contracts coincidentally ended shortly after the so-called Dubai Ports Deal controversy. For example, not only did the Indianapolis Airport Authority end its management contract in 2007, that same year the PNYNJ bought out a foreign corporation's lease on Stewart Airport.[37]

The fact that US airports are locally owned and managed is important to understand. Local and private monies provide much of the capital needed for airport expansion and improvement projects. However, the federal government is also an important player. As noted, Congress's action to deregulate the airline industry had profound impacts on the nation's airports. Federal funding has also proved crucial. As matching funds or outright grants, federal monies have also financed many of the changes and improvements at US airports since World War II (WWII).

A Reluctant Partner: The Federal Aid to Airports Program (1946–1970)

The federal government had provided some aid to local governments for airport development as early as the 1930s, and this aid expanded dramatically during WWII, though in both cases it was seen as temporary and emergency in nature. As the war ended, local governments and aviation boosters pressed for a permanent program of federal aid to airports. Airports were, after all, part of a national system of civil air transportation. Supporters also noted that airports would play an important role in any national system of air defense. And there was a fairly widespread perception that many Americans would begin to use private aircraft in very large numbers. Therefore, there was an immediate need for all types of airports.

The debates over a federal airport aid program involved a number of important issues. Much, of course, focused on the amount of federal aid that should be made available and to which airports should it go. However, before any aid could flow, a more fundamental issue had to be settled. Would federal airport aid go directly to the owners and operators of the airports (mostly municipal governments) or should aid go to the states that would then allocate funding to the various projects within their jurisdictions. Traditionally, states mediated any interaction between the federal government and cities. As the airport legislation worked its way through congressional committees, states argued that funds for airport development should flow through states to the cities. State officials made this argument despite the fact that with very few exceptions they had not played much of a role in the development of airports. Nonetheless, state officials argued that, similar to federal aid for highways, monies coming from the federal government should go to states and then to cities. City officials, not surprisingly, argued that the federal aid should come directly to them as they traditionally had been the sponsors, owners, and managers of the nation's airports. Further, Donoh W. Hanks, Jr., the Washington representative of the American Municipal Association, asserted that airport development was just one of a number of areas—housing and public works being others—where states had ignored the needs of cities. In the absence of state aid, cities had been forced to turn to the federal government. He concluded, however, that regardless how aid reached cities—directly from the federal government or through the states—aid was needed.[38]

The final bill, passed after more than a year of debate, compromised on the issue of direct versus indirect aid to cities for airport development. In late 1945, two different versions were still making their way through congress, and both had some provision for aid going through states to cities. The Senate bill directed the aid go through the states in those cases where the states had aviation agencies. Only in cases where a state did not establish an aviation agency to manage the funds could cities apply directly for aid. The House bill also directed aid go through state agencies where they existed, but further allowed any local airport sponsor to apply directly to the Civil Aeronautics Administration (CAA) for a portion of the funds allocated to the state in which they were located, unless state law specifically prohibited such an action. Though Senator Owen Brewster (R-Maine) continued to oppose any legislation that would allow any form of direct aid to cities, Senator Pat McCarran (D-Nevada) guided the bills through a conference committee that accepted the House version. The House passed the bill in April and the Senate in May 1946.[39] Both Brewster and McCarran were known for their involvement in aviation-related activity in Congress as Brewster led a controversial investigation of Howard Hughes over wartime aircraft contracts and McCarran had helped push through the Civil Aeronautics Act of 1938 and campaigned for the creation of an independent US air force. Both were also staunch anti-communists and allies of Senator Joseph McCarthy.

The issue of which airports—large or small or both—should receive federal aid also contributed to the lag between the initial introduction of a bill and final passage. As the process of developing a post-war airport aid program began, Clarence Lea (D-California), a 15-term congressman in his final term, developed his own version of an airport aid bill. Lea requested a program proposal from the CAA and it responded with an ambitious 10-year, $1.25 billion proposal, calling for the construction of thousands of airports, large and small, throughout the country. The cost would be shared 50–50 between states and the federal government.

Half of the funding would go to small airports (Class I, II, III)[40] and half would go to the larger commercial airports (Class IV, V).[41]

The emphasis on smaller airports and private flying reflected the mindset of CAA Administrator Theodore P. Wright, who served in that position from 1944 to 1948. He was a strong believer in what historian Joseph Corn has called "the winged gospel," a collection of beliefs concerning the airplane and its potential beneficial role in shaping American society and culture. One of the strongest and more enduring tenets of the winged gospel held that piloting an airplane might one day become as common as driving an automobile. Wright envisioned a post-war nation in which Americans would buy and fly as many as a half-million small airplanes, thus enjoying an unprecedented freedom of movement. A dramatic spike in small aircraft sales in 1946 seemed to portend just such a future. However, it was followed by an equally dramatic (and sustained) drop in the sales of such aircraft. Regardless, Wright continued to believe that if aircraft manufacturers could produce a small, affordable, and safe aircraft Americans would indeed take to the skies and that belief helped shape the federal airport aid legislation.[42]

The Federal Airport Aid Act favored small airports in a number of ways. First, for the smaller Class I, II, or III airports, the federal match was 50 %. For the larger Class IV and V airports, however, the match could be much less. For projects up to $5 million, the 50 % match was available. However, in the case of projects costing more than $5 million, the federal match decreased by 10 % for every $1 million in additional costs up to $11 million. Therefore, for large airport projects costing $11 million or more, the federal match would be only 20 %. The CAA Administrator, however, could provide some additional funding to the larger airport projects as the legislation placed 25 % of the total aid in a discretionary fund controlled by the administrator. (The other 75 % of the aid went to states based on a population/land area formula.) Further, the legislation required that in the case of the larger projects, the CAA administrator had to submit to Congress a report no later than two months before the end of the fiscal year detailing the funding requests for Class IV and V airports to gain the authorization to undertake them in the next fiscal year. The final passage of the act late in the fiscal year did not give the CAA the time to submit such a report and that meant all of the first years' aid went to the smaller airports.[43]

Though the very first allocations made under the airport aid program went exclusively to smaller, general aviation airports, within a few years, attention and funding shifted to those airports handling commercial air service, especially the very largest airports. Several factors played into the shift. First, Theodore P. Wright left as administrator of the CAA on March 1, 1948. Second, the outbreak of the Cold War and then the Korean War renewed the emphasis on larger airports and the role they played in the national defense. Though the total number of projects involving larger airports was small, the amounts allocated were far larger than those for general aviation airports. By 1950, for example, almost 70 % of the money allocated went to larger airports. This reflected a shift not only in federal priorities but also in state and local priorities. As noted, 75 % of the monies each year were distributed to the states by a fixed formula, and the CAA administrator distributed 25 % from a discretionary fund. Beyond distributions from the discretionary fund, in order for more federal aid to go to larger airports, the states would have had to have placed greater priority on larger airports, and both state and local governments would have had to have raised more in matching funds. By fiscal 1951, most all of the substantial grants went to airports serving some of the nation's largest cities, including Boston, Los Angeles, New York, Philadelphia, and San Francisco, as well as to some of the nation's fastest growing cities, including Houston, Miami, and Phoenix.[44]

While the increased emphasis on larger airports no doubt pleased the nation's big city mayors, it came within a context of shrinking federal appropriations. The legislation created a seven-year, $500 million program, authorizing spending of approximately $71 million annually, although within the overall ceiling, the Congress could appropriate up to $100 million in any given year. Throughout the first decade of the original program—extended by an additional five years in 1950–1958, but without an increase in the total amount authorized—Congress repeatedly failed to appropriate the amount authorized. In the first fiscal year, 1947, Congress appropriated $47.75 million. The following year the amount dropped to $30.4 million. Spending increased modestly in 1949 and 1950, but fell again with the Korean War. By 1953, the appropriation fell to only $10.2 million, and the program was suspended for all of 1954 pending an extensive review. In contrast to spending through the federal airport aid program, defense spending during the Korean War included $500 million for airports with military usefulness, thus also clearly of national importance.[45]

The review and suspension of the program in 1954 came in the wake of widespread criticism as well as more general efforts to cut the federal budget. In 1953, as Congress worked on the budget for fiscal year 1954, the federal airport aid program came under intense fire from a number of Republicans, now in the majority, many of whom questioned the very need for a program. For example, the House Appropriations Committee investigated the program and found "waste and questionable administrative practices at nine specific airports." The investigation coincided with the Department of Commerce's decision not to seek further funding until completion of a thorough program review. Despite the House findings, others on Capitol Hill fought the elimination of funding for 1954 citing complaints from cities that had planned constructions projects—and in some cases had raised the local share—based on anticipated federal matching funds. In the end, despite efforts in both the Senate and the House to include funding, the 1954 fiscal year budget eliminated all funding as Congress debated revising the program.[46]

In part responding to intense pressure from a number of aviation constituencies, including the nation's mayors, in April 1954, Undersecretary of Commerce for Transportation Robert Murray presented a $33 million airport aid program. The one-year program aimed at not only renewing the federal aid program, but also revising the program significantly. Murray argued that the previous program represented "an unsound investment of federal funds." He asserted that any renewed program had to be large enough to address national needs, had to focus on the most pressing projects, and not spread the funding out over a large number of projects, thus limiting the amount available for any given project. He found that the previous program, especially in 1953 when only a little over $10 million had been appropriated, was not selective and tried to fund too many projects, diluting any impact it might have. He also stated that a federal airport aid program clearly must address national rather than local priorities and, thus, meet the needs of a national air transportation system. Finally, his proposed program eliminated money for terminal buildings. These he saw as revenue producing and believed that they should be funded locally.[47]

Murray's proposal immediately ran into intense resistance from Senator Pat McCarran as he objected strongly to a number of provisions in the proposed legislation. First, he opposed increasing the discretionary fund from 25 % to 50 %. As he viewed it, that measure would increase the Commerce Department's control over allocating funding. McCarran, in fact, charged that Murray wanted "to become the 'czar' of the airport program." Second, McCarran objected to the requirement that approved projects had to be deemed of national, rather than strictly local, importance and that if a state proposed no projects deemed of national significance, its share of the federal funds could be diverted to other projects. Finally, he opposed the ban on the use of federal funds for terminal construction. In addition to McCarran, the proposed legislation also ran into resistance from a number of House Republicans. They claimed that the proposed legislation violated the original intent of the act, which had favored smaller airports. They saw the legislation as being designed to benefit big city airports, denying aid to the smaller airports.[48]

Legislation in 1954 reinstated federal airport aid, but it did so for only one year. Following on the heels of that measure, Congress worked to revive the program for the longer term. In July 1955, the Senate (unanimously) and the House passed a bill that created a four-year (fiscal years 1956–1959), $63 million per year airport aid program. Senator A.S. Mike Monroney (D-Oklahoma) introduced the bill. After McCarran's death in 1954, Monroney emerged as the strongest voice for civil aviation in Congress.[49]

By the time that the four-year extension was set to expire, Congress directed a major reorganization of the CAA. In 1958, Senator Monroney introduced legislation to transfer the functions of the CAA, which operated under the Department of Commerce, to a new independent agency. Congress approved the legislation and President Eisenhower signed it into law on August 26, 1958. Eisenhower appointed retired US Air Force (USAF) General Elwood "Pete" Quesada as the first administrator of the newly established FAA. It began operations in late 1958.[50]

One of the first challenges facing Elwood Quesada was a renewed fight over the federal airport aid program. In 1958, Senator Monroney sponsored a bill to extend and expand the program beyond the end of Fiscal Year (FY) 1959. His proposed legislation called for a continuation of the program through FY 1963 and an increase in the annual authorization to $100 million. Even though the legislation passed both the Senate and the House by substantial margins, President Eisenhower used a pocket veto to kill it. Eisenhower's opposition to the expanded program reflected his small government philosophy. He wanted the government out of the airport aid business, calling instead for a four-year program that would allow for an "orderly withdrawal" of the federal government from the aid program. Monroney responded by immediately resubmitting legislation at the beginning of the 1959 legislative session. His new bill called for a five-year, $575 million program ($100 million annually for five years, plus an emergency $75 million accelerate action under the new program). The administration, led by Quesada, countered with a four-year, $200 million plan aimed at closing out the program. Additionally, the administration bill called for an increase in the discretionary fund from 25 % to 50 %, and it proposed eliminating any funding for terminal construction.[51]

Quesada was the only witness to oppose the Senate bill, while many airport constituencies supported the Monroney legislation and voiced strong opposition to the administration proposal. For example, Mayor William Hartsfield of Atlanta spoke on behalf of the US Conference of Mayors. He saw the administration's alternative proposal as "woefully inadequate."[52] Despite strong support and Senate willingness to trim the package to a four-year, $465 million program, Monroney failed to get new legislation passed. The Senate bill called for a much more extensive program than that approved by the House ($297 million), and the president promised another veto. In the end, all Monroney could do was help push through another two-year extension of the program at the existing funding ($63 million per year) level.[53]

Debate over the federal airport aid program continued into the 1960s, and overall budget issues remained important in determining the size of the program. By the middle of that decade, the problems facing the nation's commercial airports became acute. Not only were they struggling to cope with the issues involved in the shift from piston-engine airliners to the larger jet airliners, but passenger traffic was growing dramatically. While Congress voted another short-term extension to the existing program in 1966, more extensive, long-range changes to the quarter-century-old program were in the offing, particularly after overall budget woes due to the escalating conflict in Vietnam led to a cut in the appropriation for airport aid in FY 1967.[54]

Creating the Airport and Airway Trust Fund

As early as 1966, FAA Administrator William F. McKee, a former USAF general, called for the creation of a task force to examine the problems with the nation's airports. The proposal went nowhere until the chair of the CAB, Charles S. Murphy, similarly called for action. Murphy, a lawyer and confidant of former President Harry S. Truman, viewed those problems as a great threat to the national air transportation system. He called for the creation of a government-industry task force to thoroughly investigate the situation and make recommendations for adequately funding needed improvements. President Lyndon Johnson responded with the creation of an Airports Task Force, chaired by Murphy and Secretary of

Transportation Alan Boyd. Johnson's sense of urgency may have been heightened by the fact that a congressional subcommittee, with information from the FAA, was also studying the issue with an eye to increasing funding to the nation's smaller airports. That subcommittee released its findings in January 1967 as the presidential task force began its work.[55]

While the congressional report focused on the nation's smaller airports, the presidential task force looked at several types of airports needing aid, including large commercial, small commercial, and reliever airports as FAA reports indicated all three types of facilities needed additional funding. The task force believed that the larger commercial airports were capable of generating sufficient revenue to finance most of the needed improvements. The problems they faced in executing improvement plans were local in nature—resistance to increased aircraft noise, limits to municipal indebtedness, and the high cost of land in large urban areas. The task force suggested a combination loan and loan-guarantee program to overcome at least some of the local financial issues. As a last resort, if local obstacles could not be overcome, the task force recommended that the federal government move in to build and operate the needed facilities. The task force also called for a loan program for the smaller commercial airports. Only airports receiving subsidized air service or those serving as general aviation reliever airports would be eligible for federal grants. A proposed Federal Airport Corporation would administer all loans. User fees, paid into a new trust fund, would finance the grant program.[56]

Even before the report went to the president, industry groups voiced their strong opposition. The Airport Operators Council International (AOCI),[57] the organization representing commercial airport owners, called for an expanded federal grants program of $300 to $400 million per year. Along with the ATA, representing the nation's airlines, the AOCI objected to the exclusion of terminal facilities from the program. And the Aircraft Owners and Pilots Association, representing general aviation interests, saw the new program as primarily benefiting the larger commercial airports. It also strongly opposed the idea of user fees to pay for the program. Once presented to the White House in May 1967, the report also found disfavor with members of the president's staff. Their opposition was based on the growing feeling that the airlines must bear a larger share of the cost of improving the system.[58]

About two months after the task force presented its findings to the White House, the House Commerce Committee scheduled hearings. The hearings, however, were not focused narrowly on airport funding.

Instead, as a result of two spectacular mid-air collisions—one in Ohio and a second in North Carolina—the issue was aviation safety. The committee hearings, though, tied the safety issue to the problems facing the nation's airports. FAA Administrator McKee agreed with the committee members that safety and airports were linked. The White House disagreed and was concerned that the recent airport task force's report—which had not been released to the public—might somehow be labeled as an examination of aviation safety and the failure to release it an attempt to cover up problems. Despite administration attempts to decouple the issues, airports and safety were further joined when Senator Monroney's aviation subcommittee also began hearings on problems in the nation's airport and airways system.[59]

The Senate hearings did not bring up the airport task force report as the White House feared. However, they did expose divergent policy views within the administration. At the Senate hearings, Secretary of Transportation Boyd called the current airport aid program inadequate, but he did not see a great deal of support for expanding it. He did not believe that the problem with airports rated as a high priority at any level of government—local, state, or national. Any revised program, however, would need additional capital, and Boyd believed that had to come from user fees. CAB Chair Murphy, on the other hand, testified that the development of the nation's airports was a high priority, particularly for the national government. Airports were part of the national air transportation system. Central planning and funding were needed to create a system that met national needs.[60]

The hearing also revealed some disagreements between congressional leaders and municipal officials. Sam Massell, Jr., the vice mayor of Atlanta, came to Washington to represent both the National League of Cities and the US Conference of Mayors. He noted that Atlanta had recently constructed a new $20 million terminal at its airport. However, in the six years since its completion, traffic had grown so dramatically at Hartsfield International that plans were already in the works for a $50 million terminal project. The city was still paying off the debt from the original project and did not have the funds to undertake the new project. Mansell called for the implementation of a user or passenger fee to help fund airport improvements. Monroney countered Mansell, reminding him that the current federal airport aid program did not allow funding for terminal projects and noted that the airlines and other witnesses at the hearings objected to any user or passenger fee. Monroney also questioned whether large and prosperous airports, such as Atlanta, needed federal aid and suggested that the airlines could provide more funding for airport improvements. Monroney viewed the federal aid program primarily as addressing safety, thus monies should go to runways and navigation aids, not terminals.[61]

As a consequence of the attention the congressional hearings drew to the safety issue, in September 1967, President Johnson charged Transportation Secretary Boyd with developing long-range plans to improve the nation's airways and air traffic system. Boyd delegated responsibility for the requested proposal to the FAA. There the issue again expanded to include airports as the FAA viewed those facilities as part of the overall air traffic system. As work progressed on the new report, it became apparent that the FAA and the office of the Secretary of Transportation disagreed on a number of issues, especially on that of a trust fund to finance airport grants and the expansion of the grant program itself, both of which the FAA favored and the secretary opposed.[62] Further complicating the situation, the preparation of the DOT report overlapped the preparation and issuance in early 1968 of a congressional report on airport congestion by Moroney's aviation subcommittee. Despite the reservations Monroney expressed during the hearings, the report called for a broad revision and expansion of the federal airport aid program to include increased funding for all types of airports, implementation of user fees, and the creation of an aviation trust fund.[63]

Finally, despite strong objections from the FAA, the DOT submitted its own version of an airport and airways modernization plan to Congress in May 1968. The proposal came in the form of two separate bills—one dealing with the airways and the second with airports. The two bills, especially the airport bill, met with an immediate negative reaction from Congress, especially from Senator Monroney. The administration's proposal called for a federal loan program for the larger airports and a limited grant program for some of the smaller commercial airports with subsidized service. Secretary Boyd defended the loan program stating that the nation's largest airports no longer needed federal grants. He believed that local user fees—in this case landing fees, not passenger fees—paid by direct users of the airports could finance any needed improvements. Monroney rejected the secretary's argument. Although he had earlier suggested that the airlines might contribute more to airport expansion projects, he did not see the airline industry, which would pay the landing fees, as being "in a state of financial strength to bear the total cost and the total responsibility for national airports improvements."[64] In contrast, Moroney called for a ten-year, $1 billion program. Instead of landing fees, he called for broader-based user fees or taxes on both commercial and general aviation. The funds raised would go into an aviation trust fund to be used exclusively for airport and airways development and maintenance. Though work continued on legislation, mostly in the form of Moroney's bill, in the end Congress failed to act in 1968. Revision of the airport grant program would be left to the incoming Nixon Administration.[65]

As soon as the 91st Congress began its new session in early 1969, a number of congressmen introduced bills to expand aid to airports. Although Senator Monroney had lost his bid for re-election, his 1968 bill appealed to the new administration. The acting FAA administrator, David D. Thomas, recommended that the administration base its proposal on the Monroney bill. New Secretary of Transportation John A. Volpe, former governor of Massachusetts and the first head of the Federal Highway Administration, also supported the trust fund-based proposal. Aviation organizations, however, continued to express misgivings. Basically, many did not trust the DOT, especially general aviation groups that anticipated heavy taxation under the trust fund plan. Further, all aviation groups feared that, despite promises, funds raised for the trust fund could and would be diverted for other uses. To overcome their objections and gain support for the legislation, the DOT officials worked to sell the idea of user taxes to the aviation community.[66]

By the time the administration's bill was introduced and the hearings underway, aviation groups had come around to a position where they basically supported the administration's proposal. Most of the debate concerned exactly what the user taxes would be and how the tax burden would be distributed over the range of aviation interests. The main issue was how the commercial airlines' contribution would be made—passenger ticket taxes and cargo taxes versus a fuel tax. The commercial airlines favored the former; general aviation interests pushed for the latter. Finally, another major issue involved whether or not the trust fund would support airport terminal projects. As the process went forward, both the ATA and the AOCI pushed for the inclusion of funding for airport terminals. While Senate supporters sided with those two groups, House members insisted on removing the funding. In the end, the Senate agreed to drop funds for terminals, while the House agreed to five-year, rather than a three-year authorization. The Airport and Airway Development and Revenue Acts of 1970 passed overwhelmingly in both houses of Congress and President Nixon signed them into law on May 21, 1970.[67]

A Not So Stable Source of Funding: The Aviation Trust Fund, 1970–1982

The Airport and Airway Development and Revenue Acts of 1970 created the Airport and Airway Trust Fund, also known as the Aviation Trust Fund. Monies for the trust fund came from two existing taxes (on aviation gasoline and the on passenger tickets on domestic flights) and three new taxes or fees (taxes on international passenger tickets and air freight way bills, and an annual aircraft registration fee).[68] The formula for distributing aid—75 % to the states and 25 % discretionary—established under the previous airport aid program basically continued. The purpose of the legislation had been to provide both increased and stable funding for the development of the nation's airports and airways, and in some ways the program initially met its promise. During the first five-year authorization period, the federal government obligated $1.3 billion for airport improvement and $37.5 million for planning. During the 24 years of the previous Federal Aid to Airport Program, only $1.2 billion had been obligated for airport improvement and nothing for planning.[69] However, a number of controversies raged that resulted in significant amendments to the program during the first few years.

The Airport and Airway Trust Fund was not even a year old when the Nixon Administration proposed a major restructuring. This involved not just aviation funding, but federal funding for all transportation projects. The administration proposed creating a single transportation trust fund. Out of that single fund, states would receive block grants that they could spend as they saw fit. Transportation was not the only area to see the shift toward less restrictive block grants. As part of what he termed his "new federalism," for example, President Nixon pushed for the replacement of most specific urban aid programs with so-called community development block grants.[70] Aviation interest groups that had grudgingly been led to support the aviation trust fund idea strongly opposed the idea of a combined transportation fund. Even though Secretary of Transportation Volpe sought to assure them that the monies paid into the new trust fund from aviation user taxes and fees would be used to support aviation projects, the proposed legislation contained no such specific language. Aviation interests strongly objected to any possible diversion of the monies they paid into the trust fund going to non-aviation projects.[71] In the face of strong opposition, the plan for a unified transportation fund faded. Meanwhile, in November 1971, Congress passed an amendment to the legislation aimed at assuring the aviation community that the trust fund monies would be used as originally promised. The amendment prohibited the diversion of trust fund monies to FAA operations, required any surplus funds to be used exclusively for the administration of the airport and airways development program, and stipulated that all funds authorized each year remain available until obligated.[72]

Further, in June 1973, Congress passed and President Nixon signed the Airport Development Acceleration Act. This act further amended the original legislation. Given the lag in obligating funding during the first years of the program, the new law increased the annual authorization for airport aid to $310 million per year. In addition, it raised the federal share for projects at smaller airports from 50 % to 75 %. The legislation also prohibited local or state governments from levying a "head tax" on airline passengers, which had been proposed to help states and local governments raise their share of matching funds.[73] The issue of the "head tax" would recur over the next decades.

The Airport and Airways Development Acts' original five-year authorization was scheduled to lapse at the end of FY 1975. As the program came up for renewal, the Ford Administration sought several major changes. Most controversially, despite the 1971 measure specifically prohibiting such an action, it called for using $431 million from the trust fund for FAA operations and maintenance in FY 1976. Aviation supporters in Congress, however, had other ideas that involved expanding the grant-in-aid program in a number of ways. First, the lag between appropriations and authorizations had created a surplus. While the Ford Administration sought to use some of the surplus to fund FAA operations and maintenance, aviation boosters sought instead to increase spending on airports. Second, they also sought to broaden the types of projects that could benefit from the grants program including terminal improvements and land acquisition around airports to address the jet aircraft noise issue. Finally, supporters called for increasing the federal share in all airport projects—to 75 % at larger airports and to 90 % at smaller airports. By April 1975, it was clear that the administration's proposal had little or no support in Congress. However, it was also clear that Congress was unlikely to put together its own legislation before the program lapsed at the end of June.[74]

In fact, negotiations between the House and the Senate continued well into FY 1976, and it was not until June 1976 that they concluded. In the end, the House and Senate agreed to a five-year $5.7 billion extension of the program through FY 1980. Under the legislation, the amounts authorized for the grants-in-aid program increased incrementally from $500 million in FY 1976 to $610 million in FY 1980. It also reversed the 1971 action and allowed diversion of $250 million from the trust fund beginning in FY 1977 for FAA operations and maintenance and the amount that could be diverted increased by $25 million per year through FY 1980. The new law also increased the federal share of projects at larger commercial airports to 75 % and at smaller commercial airports to 90 %.[75] And it expanded the types of projects eligible for federal aid to include, importantly, such noise-abatement projects as landscaping, sound barriers, and the purchase of land around airports to serve as a noise buffer.[76]

Despite the increased spending, the surplus in the aviation trust fund continued to grow throughout the late 1970s as actual appropriations continued to lag the authorizations. As budget deficits grew, many began to charge that the federal government was using the surpluses in various trust funds—social security and the highway trust funds as well as the aviation trust fund—to mask the true size of the federal budget deficit.[77] Authorization for the aviation trust fund was scheduled to lapse on June 30, 1980. As Congress and the Carter Administration looked to replace and revise the existing program, the issue of the trust fund surplus played a major role.

Deregulation, Defederalization, and the Airport Improvement Program

As the closing date for the aviation trust fund neared, Congress and the Carter Administration contemplated changes to the airport grant program. The late 1970s and early 1980s were a particularly difficult time for the aviation industry. As noted, President Jimmy Carter signed the Airline Deregulation Act in 1978, aimed at lowering the cost of flying. The first several years of deregulation proved a period of fierce competition, price cutting, and uncertainty. Adding to the complexity, as noted, the start of deregulation coincided with the second oil shock, a deep recession, and President Reagan's decision to fire all striking air traffic controllers.[78] Amidst all this, Senator Howard Cannon (D-Nev.), a USAF reserve officer and pilot who succeeded Mulroney as the primary voice for aviation interests in Congress, led a campaign to drastically reshape the federal airport aid program. In the end, the revisions were not as radical as originally envisioned. Nonetheless, several significant changes were made.

Anticipating the expiration of the existing airport aid legislation, Senator Cannon introduced a bill in September 1979 to replace the Airport and Airways Development Acts of 1970. His bill would direct that the user fees collected at the nation's largest airports stay at those airports rather than going to the aviation trust fund. Further, the ticket tax, a major source of funding for the federal airport aid program, would be reduced from 8 % to 2 %. Only smaller airports would continue to receive federal aid financed by these lower taxes. Cannon's argument was that since the Office and Management and Budget refused to allow appropriations from the trust fund to match congressional authorizations—both to deal with inflation and to use the surplus to offset the federal deficit—it would be better to have those funds simply stay at the larger airports where they were collected. Those larger airports could then use the funds for needed capital improvements. Further, those airports would no longer be part of the federal airport aid program—they would be "defederalized."[79]

That same month FAA administrator Langhorn M. Boyd, the son of a pioneer aviator and airline executive, testified before Congress. He told the Senate that the nation's largest airports could survive without federal aid, though in many cases it might involve hiking fees for airlines and, in a few cases, relying on local support. While he thought the Cannon proposal was "innovative," he also stated that he could not support it. Several aviation groups, on the other hand, testified with highly qualified support for the Cannon bill. Essentially, they approved of the bill generally, but were concerned about many of the details. *Aviation Week and Space Technology*, a major voice for aviation and aerospace interests, published an editorial on the Cannon measure in October 1979. Essentially, it acknowledged the problems the Cannon bill aimed at addressing—not just the use of the surplus to offset the deficit, but the continued use of the trust fund to also provide a significant part of the FAA's budget. The editorial, too, was very cautious in its approach. While not rejecting the proposal, it expressed concerns about many of the details, including excluding the largest airports from the federal aid program and allowing airports to use a passenger head tax to replace federal monies.[80]

By the end of 1979, opposition to the Cannon measure began to mount. Though some individual airport operators continued to favor the proposal, the AOCI asked for modifications. First, the AOCI proposed that large and medium commercial airports be allowed to voluntarily leave or stay within the program. Further, if an airport did decide to defederalize, it should be allowed to impose a passenger facility charge—a new name for the passenger head tax specifically prohibited under a 1973 amendment to the trust fund legislation. The organization's executive vice president, J. Donald Reilly, also questioned just how self-sufficient many of the larger airports actually were and to what degree they would be able to negotiate higher fees from the airlines. He noted that some airport operators predicted that without federal funds they would be completely dependent upon the airlines for needed capital. And the airlines, struggling to adjust to the new world of deregulation, were not likely to be willing to negotiate higher fees. Finally, a number of airport operators also noted that the lack of predictable federal funding would make it harder for airports to issue bonds to cover projects.[81]

Debate over the Carter Administration proposal, the Cannon bill, and other proposals in the House of Representatives continued into 1980. Even as the deadline loomed for the expiration of the program, no compromise was found. As a result, the enabling legislation lapsed and several of the user taxes earmarked for the trust fund expired or were decreased and the funds that were collected went to the general treasury rather than the trust fund. Projects already approved continued to receive funding, as the surplus in the trust fund provided needed monies, but newly proposed projects would have to await congressional action on renewal legislation. Additionally, the airport community was divided over the issue of defederalization. While a few large airports still supported the idea, most opposed. On the other hand, many smaller airports approved of the proposal seeing it as a way to gain a larger share of the federal aid funding. By fall 1980, it was clear that any renewal legislation was unlikely to pass until after the election and Senator Cannon announced he would block any legislation that did not include his defederalization plan.[82]

In March 1981, Senator Nancy Kassenbaum (R-Kansas) introduced new airport aid legislation. It was similar to the Cannon bill, but would allow airports that defederalized to levy a passenger facility charge. Under the new version of the legislation, 69 large airports would be defederalized by September 30, 1981. An additional 29 would be out of the program by September 30, 1982. The new Reagan Administration came out in favor of the Senate plan for defederalization, but proposed its own bill that would defederalize only 41 airports and raise user fees. Both the Kassenbaum measure and the Administration's bill differed greatly from the proposed measure in the House which rejected defederalization. Representative Norman Mineta (D-Calif.), chair of the House's aviation subcommittee and future Secretary of Transportation, viewed defederalization as "the single greatest obstacle" to new airport aid legislation. The administration plan also faced opposition from Senate Republicans who objected to the increased user fees and the smaller number of airports defederalized. Though both the House and the Senate reported out legislation by May 1981, the vast differences in the bills again did not yield to compromise.[83]

By 1982, the lack of new legislation became critical. The FAA was predicting that a number of the nation's larger airports would reach their capacity for handling air traffic within the next decade. In the absence of expansion and improvements, capacity restrictions might be necessary. Airlines, while favoring defederalization, opposed the passenger facility charge. Airport operators, who by 1981 by and large had come to oppose defederalization, were willing to give up on the passenger facility charge and to have more of the FAA budget funded out of the trust fund. In return, though, they wanted an increased federal aid program (from $450 million per year to $700 million).[84]

A new federal aid to airports program finally passed as part of a larger tax bill, the Tax Equity and Fiscal Responsibility Act, in September 1982. Title V of the new legislation renewed the Aviation Trust Fund and reinstated and/or raised the several aviation user fees and taxes funding it. It also added a 14 cent per gallon tax on jet fuel. The new airport aid program, the Airport Improvement Program (AIP), was authorized funding through FY 1987. The amount authorized increased from $450 million in FY 1982 to $1.017 billion in FYs 1986 and 1987. The formula for distributing airport aid was revised, but more importantly instead of removing airports from the federal aid program, the new legislation made privately owned, public access commercial or reliever airports eligible for federal aid. Additionally, a separate Surface Transportation Assistance Act of 1982 authorized $450 million for airport aid between FY 1983 and 1985. Then in 1987, President Reagan signed new legislation, the Airport and Airway Safety and Capacity Expansion Act. It renewed the AIP through FY 1992.[85] Authorizations continued to outpace appropriations, however, and the surplus in the trust fund again emerged as a major issue again in the 1990s.

Airports for the Twenty-First Century

The most recent revision of the federal airport aid program came after a decade of continued debate over the program. During the 1990s, the issues included the relationship between the trust fund surplus and the federal deficit, capacity issues at many major airports, overall reform of the

FAA, and the use of the trust fund to finance the FAA budget. In many cases, the federal aid program itself was not the issue. Rather, it was caught up in controversies surrounding those related issues. In the end, Congress voted to increase spending while tightening federal rules governing airports receiving monies from the aviation trust fund.

Airport delays and issues with capacity were not new concerns. However, with the growth of air travel in the wake of deregulation, they had high visibility by the late 1980s and early 1990s. In 1990, airport supporters proposed that Congress increase the amount of money available for airport grants by hiking user fees and allow airports to implement passenger facilities charges (PFC) to finance capital improvements not covered by the federal grants. While the administration of President George H.W. Bush supported those proposals, Democrats in the House, led by Congressman Norman Mineta, strongly opposed. He, as well as general and business aviation groups, opposed any increased user fees. Both argued that even under the current structure, the trust fund continued to build up a surplus. Unless the administration was willing to support greatly increased airport aid spending, Mineta saw no need for higher fees. He also suggested that if the larger airports were allowed to charge the PFCs, such an action should also make them ineligible for federal funding.[86] The eventual legislation—the Omnibus Federal Budget Reconciliation Act of 1990—did not increase airport aid authorizations over those already approved through FY 1992. However, it did authorize the FAA to set rules under which airports could begin to implement PFCs without losing access to federal airport grants. The initial maximum amount was set at $3.00 per passenger and the first airports to begin collecting the PFCs in May 1991.[87]

From FY 1992 through FY 1999, battles over a host of issues resulted in several short-term extensions of the program. The exact issues varied from year to year, but generally involved efforts to cut federal spending and reform the FAA. Several times Congress missed deadlines to extend the authorization not just for spending, but for the collection of the user fees and taxes that supplied the trust fund. For example, the grant program lapsed between June 1994 and August 1994. Authority to collect the aviation user fees and taxes expired on December 31, 1995. It was not renewed until August 1996, but then expired again at the end of FY 1996. In September 1996, President Clinton signed a bill that extended the grant authorizations until the end of FY 1998, but the excise tax on passenger tickets was allowed to lapse for several months.[88]

Many of the lapses in spending and collection were due to the larger budget battles waged during the 1990s. In addition, in the mid-1990s, Senators John McCain (R-AR) and Wendell Ford (D-KY) led an effort to reform the FAA, to make it more efficient and cost-effective. They pushed for legislation that would give the FAA a certain level of independence on matters such as personnel management, procurement, rules making, and budgeting. They also envisioned an FAA funded out of user taxes and new fees, which would go directly to the FAA rather than through the aviation trust fund. Although the new fees aimed at paying the cost of certain FAA services, such as air traffic control, would not be implemented unless the existing taxes were lowered, the new fees would result in charges for services that were previously provided free. General aviation interests, in particular, strongly opposed any of the new user fees, arguing that they would hurt aviation safety as many general aviation pilots would simply not use FAA services, such as weather briefings and information for flight planning, if they involved fees.[89]

The FAA reform initiative was also part of a larger effort on the part of the administration of President William J. Clinton to cut $6.9 billion from the DOT's budget. In early 1995, the Clinton Administration proposed a number of spending cuts, including federal grants to small reliever airports. The administration also suggested that the larger airports authorized to collect PFCs be dropped from the federal aid program. Edward M. Scott, a spokesman for the National Association of State Aviation Officials, in testimony before Congress objected to the proposed cuts in funding to smaller airports, but supported defederalizing larger airports. He suggested that if thc administration raised the $3.00 cap on the PFCs, "cutting loose" those larger airports from the federal program would be possible.[90] The final legislation was far more modest than had been proposed. It did reform the FAA's personnel and procurement systems, but most of the other provisions, including the new user fees and the greater independence for the FAA, were eliminated from the final bill. The AIP, however, was further cut. Spending from the grant program had reached $1.8 billion in FY 1993. It dropped to $1.7 billion in FY 1994 and $1.4 billion in FY 1995. Grants awarded dropped further to $1.3 billion in FY 1996.[91] On the other hand, Congress did come to some consensus over the taxes that fed the trust fund. The Taxpayers Relief Act of 1997 authorized the collection of the aviation taxes paid into the trust fund through 2007. It also outlined when and how such taxes could be increased and added a new international arrival tax. While authorizations for spending the monies continued to experience periodic lapses, at least through 2007 the taxes themselves did not lapse, as they had on other occasions.[92]

Though Congress found agreement on the collection of the taxes, a number of other issues proved particularly thorny. Congressman Elmer G. "Bud" Shuster (D-PA), a strong proponent of transportation-related federal projects, pushed to take the aviation trust fund "off budget," so that the surpluses in the fund could not offset the federal deficit. At the same time, he and his House allies insisted that the FAA budget continue to come from both the trust fund and general tax revenues, as they viewed many FAA activities serving the welfare of the general public. The House also wanted a five-year authorization. On the Senate side, Senator John McCain and his allies opposed taking the trust fund off budget. The Senate also insisted on a three-year authorization. Both chambers, however, generally agreed that annual appropriations needed to match the funds collected each year by the trust fund. After protracted debate, Congress passed the Aviation Investment and Reform Act for the Twenty-First Century (later renamed to honor retired Senator Wendell Ford) and President Clinton signed it in April 2000.[93]

The Wendell Ford Aviation Investment and Reform Act for the Twenty-First Century (also known as AIR-21) had a number of important provisions. First, it authorized expenditures from the Aviation Trust Fund from FY 2000 through FY 2003, increased the amount of funding the FAA received from the trust fund through FY 2003, and it guaranteed that all revenues flowing into the fund would be used exclusively for aviation purposes through FY 2003. The act also increased authorizations for the AIP and mandated that it be funded at its full authorization. This was seen as a particular benefit to smaller airports. For larger airports, the legislation raised the cap on PFCs from $3 to $4.50, but limited the amount any single air passenger would have to pay to $18. It also increased the maximum and minimum amounts of annual funding larger airports could receive and guaranteed funding to general aviation airports.[94]

Conclusion

Local ownership and federal funding of the nation's airports provided fertile grounds for repeated political battles. Despite some significant starts and stops, generally the federal government provided monies for the development and improvement of the nation's commercial airports. It also generally supported local ownership and management, though opening the door to the possibility of privatization—a door through which few have even contemplated entering. Combined with the introduction of jet airliners, deregulation, economic development hopes, and security concerns, all these factors created the context for the development of US airports since 1945. How those battles played out and how local governments met the challenges of the jet age are the subjects of the next chapters.

Notes

1. See Bednarek, *America's Airports: Airfield Development, 1918–1947*, 15–18, 27–38, 50–63, 100–106.
2. Only two small commercial airports have been under full private ownership and management—Airborne Airpark in Ohio, owned for a time by DHL, and Branson Airport in Branson, MO, as a group of private investors opened a new private commercial airport near the resort/entertainment community in 2009. And one commercial airport is owned by a Native American tribe. However, in 2009, DHL announced it would close its operations at the Wilmington, Ohio, airport. DHL then transferred ownership to a local airport authority. Bob Driehaus, "DHL Cuts 9500 Jobs in U.S., and an Ohio Town Takes the Brunt," *New York Times*, November 10, 2008, http://www.nytimes.com/2008/11/11/business/11dhl.html (accessed June 18, 2012); Gary Huffenberger, "DHL to donate airpark, *News Journal*, January 18, 2010. http://www.wnewsj.com/main.asp?SectionID=49&SubSectionID=156&ArticleID=181879 (accessed June 18, 2012). Christine Negroni, "In Missouri, Investors Seek a Profit in Branson Airport," *New York Times*, April 21. 2009.
3. Public authorities have a long history in the USA. The precursors to such organizations date back to the early republic. However, since the 1930s, the USA has witnessed a proliferation of "government created agencies, launched to fulfill public services, that produce and often sell goods and services." These agencies are "incorporated independently of general-purpose governments, receive most of their revenue for the goods and services they produce, and acquire capital by selling bonds backed by their stream of revenue." Though these public authorities carry a variety of names—"commissions, trusts, districts, administrations, public corporations, and public benefit corporations"—all these enterprises are what is most commonly known as public authorities. Encouraged by both favorable court decisions and

federal government policies responding to the Great Depression, especially New Deal public works programs, the use of public authorities and revenue bonds expanded dramatically in the post-war period, to the extent that by the early twenty-first century "almost two-thirds of public borrowing is done by revenue producing agencies." See Gail Radford, "From Municipal Socialism to Public Authorities: Institutional Factors in the Shaping of American Public Enterprise," *Journal of American History* 90 (December 2003): 864, 867–868, 873–888. For other works dealing with the history of public authorities in the USA, see Eric H. Monkkonen, *The Local State: Public Money and American Cities* (Stanford, Ca.: Stanford University Press, 1985) and Alberta M. Sbragia, *Debt Wish: Entrepreneurial Cities, U.S. Federalism, and Economic Development* (Pittsburgh, Pa.: University of Pittsburgh Press, 1996).

4. Wendell E. Pritchett and Mark H. Rose, "Introduction: Politics and the American City, 1940–1990," *Journal of Urban History* 34 (January 2008): 210.
5. For an example of local interests defeating a regional initiative, see Louise Nelson Dyble, "The Defeat of the Golden Gate Authority: A Special District, a Council of Governments, and the Fate of Regional Planning in the San Francisco Bay Area," *The Journal of Urban History* 34 (January 2008). For an example of a failure of regional airport planning, see Janet R. Daly Bednarek, "Layer Upon Layer: Public Authorities and Airport Ownership and Management in St. Louis, 1947–1980," *Journal of Planning History* 8 (November 2009).
6. Pritchett and Rose, "Politics and the American City," 215.
7. For a more detailed discussion of airport development by park departments in general and of how and why the Minneapolis airport was developed by its park commission, see Janet R. Daly Bednarek, "The Flying Machine in the Garden: Parks and Airports, 1918–1938," *Technology and Culture* 46 (April, 2005).
8. Quoted in Charles Rhyne, *Airports and the Courts* (Washington, D.C.: National Institute of Municipal Law Officers, 1944), 41.
9. For a more detailed discussion of the municipal airport competition in the New York area, see Bednarek, *America's Airports*, 106–114.
10. Jameson Doig, *Empire on the Hudson: Entrepreneurial Vision and Political Power at the Port of New York Authority* (New York: Columbia University Press, 2001), 262–281.

11. Ibid. The information determining the top 50 airports in the USA is drawn from the FAA website. Again, airports in US territories and possessions were excluded. "CY 2006 Primary and Non-Primary Commercial Service Airports (updated 10/18/2007)" http://www.faa.gov/airports_airtraffic/airports/planning_capacity/passenger_allcargo_stats/passenger/media/cy06_primary_np_comm.pdf (accessed 29 May 2008).
12. The St. Louis Downtown Airport is located in Illinois. It is owned and operated by the Bi-State Development Agency. This information is drawn from data available on the Federal Aviation Administration's website. The title of the data set is "Airport Data (5010) & Contact Information."

    http://www.faa.gov/airports_airtraffic/airports/airport_safety/airportdata_5010/menu/index.cfm#reports (accessed June 17, 2009).
13. "Airport Data & Contact Information."
14. Robert Fairbanks, "Responding to the Airplane: Urban Rivalry, Metropolitan Regionalism, and Airport Development in Dallas, 1927–1965," in Hamilton Cravens, Alan Marcus, and David M. Katzman, eds., *Technical Knowledge in American Culture: Science, Technology, and Medicine Since the Early 1800s* (Tuscaloosa: The University of Alabama Press, 1996), 172–188.
15. The Airport Data & Contact Information was sorted by All FAA Regions, All Districts, All States, All Counties, All Cities, Public Use Facilities, and Certified (Part 139 Certification). Airports in US territories and possessions—American Samoa, Saipan, Guam, Midway, Puerto Rico, Virgin Islands—are not included in the totals, only airports in the 50 states.

    http://www.faa.gov/airports_airtraffic/airports/airport_safety/airportdata_5010/menu/index.cfm#reports (accessed June 17, 2009).

    The website for the Metropolitan Washington Airport Authority indicates that "all property"—meaning basically the two airports—was transferred to the Authority. It further states that the Federal Government continues to hold title to the lease. http://www.metwashairports.com/about_the_authority/history. (accessed June 17, 2009).
16. For more on the privatization of transportation facilities, particularly airports, see Jose A. Gomez-Ibanez and John R. Meyer, *Going*

Private: The International Experience with Transport Privatization (Washington, D.C.: The Brookings Institution Press, 1993), 211–230 and Clifford Winston, *Last Exit: Privatization and Deregulation of the U.S. Transportation System* (Washington, D.C.: The Brookings Institution Press, 2010), 76–101.

17. Gomez-Ibanez Meyer, *Going Private*, 212–222.
18. Carole A Shifrin, "U.S. Nears Transfer of National, Dulles Control," *Aviation Week & Space Technology* 122 (April 1, 1985): 31.
19. "Senator Calls for Private Sale of National, Dulles Airports," *Aviation Week & Space Technology* 124 (April 14, 1986): 43.
20. See Washington Metropolitan Airports Act of 1986, Title VI of Public Law 99–500.
21. Carl H. Lavin, "Albany Gets Private Offer to Purchase Its Airport," *New York Times*, October 9, 1989; ____. "U.S. Refuses Private Proposal To Take Over Albany Airport," *New York Times*, December 6, 1989; James Ott, "FAA Rejects Two Proposals to Privatize Albany Airport," *Aviation Week & Space Technology* 131 (December 11, 1989): 44.
22. Ibid.
23. Lockheed eventually won a contract to manage the Albany County Airport. Calvin Sims, "Lockheed Now a Leader in Airport Management," *New York Times*, March 25, 1993.
24. "Task Force Fails to Agree on Airport Privatization," *Aviation Week & Space Technology* 131 (March 26, 1990): 82; John H. Cushman, Jr., "F.A.A. Is Debating Private Airports," *New York Times*, March 1, 1990.
25. Ronald S. Lauder, "A Case for Selling N.Y.C.'s Airports," *New York Times*, May 6, 1990.
26. Gabriel Roth, "Airport Privatization," *Proceedings of the Academy of Political Science*, vol. 36, No. 3 (1987), 74–82. For Poole's argument on privatization see Robert W. Poole, Jr., "Privatization: A New Transportation Paradigm," *Annals of the American Academy of Political and Social Sciences* 553 (Sept. 1997).
27. James Ott, "Bush Order Opens Door for Airport Privatization," *Aviation Week & Space Technology* 136 (May 11, 1992): 24–25; See General Accounting Office, "Report to the Subcommittee on Aviation, Committee on Transportation and Infrastructure, House of Representatives, Airport Privatization: Issues Related to the Sale or Lease of U.S. Commercial Airports," (November 1996).

28. Steven Lee Myers, "Giuliani Suggests Privatizing of Airports," *New York Times*, December 10, 1994; "Don't Sell the Airports," *New York Times*, January 14, 1995; Diana B. Henriques, "Wall Street; Psst! Wanna Buy an Airport?" *New York Times*, January 22, 1995.
29. Ronald Smothers, "City Hall Accepts Bids to Operate Airports," *New York Times*, January 29, 2000; Eric Lipton, "Giuliani Says That He Lacks Power to Take Over Airports," *New York Times*, March 29, 2001.
30. Charles Bagli, "Bloomberg Administration and Port Authority Get Closer on Possible Land Swap Deal," *New York Times*, April 1, 2003; Michael Cooper, "City Offers Longer Airport Leases for $700 Million and More Rent," *New York Times*, October 16, 2003.
31. "BAA Will Manage Indianapolis Airport," *New York Times*, September 13, 1995; James Ott, "Indianapolis Serves as Privatization Testbed," *Aviation Week & Space Technology* 149 (December 14, 1998): 52; "Indy Airport Authority to take over management from BAA," *The Mooresville/Decatur Times*, May 5, 2007.
32. "BAA concludes sale of US airport retail management business," http://www.moodiereport.com/document.php?c_id=6&doc_id=24842 (accessed June 18, 2012); David Teaher, "Ferrovial lands BAA with final offer of £10.bn," http://www.guardian.co.uk/business/2006/jun/07/theairlineindustry.travelnews (accessed June 26, 2009).
33. Airports Worldwide, http://www.airportsworldwide.com/Why-Airports-Worldwide/Who-We-Are/ (accessed December 30, 2014).
34. Orlando Sanford International Airport Organization, http://www.orlandosanfordairport.com/organization.asp (accessed December 30, 2014); Airports Worldwide History, http://www.airportsworldwide.com/Why-Airports-Worldwide/History/ (accessed December 30, 2014).
35. David Teaher, "Ferrovial lands BAA with final offer of L10.bn," http://www.guardian.co.uk/business/2006/jun/07/theairlineindustry.travelnews/print (accessed June 26, 2009).
36. For a sense of the controversy surrounding the Dubai Ports Deal, see Tony Karon with Douglas Waller, "Who's Behind the Dubai Company in U.S. Harbors?" *Time*, February 20, 2006 http://www.time.com/time/nation/article/0,8599,1161466,00.html (accessed June 18, 2012); David E. Sanger, "Under Pressure, Dubai Company Drops Port Deal," *New York Times*, March 10, 2006, http://www.nytimes.com/2006/03/10/politics/10ports.html?pagewanted=all (accessed

June 18, 2012); Ken Belson, "Delaying of Dubai Port Deal Brings Port Authority Grief," *New York Times*, February 16, 2007 http://www.nytimes.com/2007/02/16/nyregion/16dubai.html (accessed June 18, 2012).

37. In many ways, the saga of Stewart International was not only an example of a terminated privatization experiment, but also another chapter in a long effort to provide the New York City region with additional airport capacity. In the midst of efforts to attract commercial airlines to Stewart, Governor George Pataki, a strong proponent of privatization, announced plans in 1995 to privatize the airport. Initially, the governor planned to sell the airport, which the state owned and a private company under contract to the state's Department of Transportation managed. However, under the FAA pilot privatization program announced in 1996, the state was limited to leasing the airport. Consequently, the state entered a multi-year negotiation with the FAA to identify and approve a private company to lease the airport. In 1998, the state announced a tentative agreement with National Express Group, a British transportation company with experience operating two regional airports in Great Britain. The 99-year lease between the state and National Express Group received final approval from the FAA in March 2000. National Express Group, however, decided in 2006 to put its least up for sale, citing continued difficulties in promoting commercial air service. The Port Authority board, in a move it said would help coordinate efforts to deal with air traffic congestion in the New York region, voted in 2007 to buy out the remaining years of the lease and took over operations in late 2007. "Pataki Proposes Sale of Stewart Airport," *New York Times*, February 18, 1996; Raymond Hernandez, "Plan to Lease Stewart Airport is Approved," *New York Times*, November 27, 1997; "Metro New Briefs: New York; British Company Chosen to Run Stewart Airport," *New York Times*, April 9, 1998; "Metro News Briefs; New York; Final Approval for Airport Privatization," *New York Times*, March 22, 2000; Patrick McGeehan, "North of the City, a Minor Airport Is on the Brink of the Majors," *New York Times*, October 25, 2006; ______., "4th Major Hub for Air Traffic Moves Ahead," *New York Times*, January 25, 2007; John Holl, "The Week; Port Authority Can Buy Lease of Stewart Airport," *New York Times*, May 6, 2007.
38. Committee on Interstate and Foreign Commerce, *Hearings Before the Committee on Interstate and Foreign Commerce, House of*

Representatives, Seventy-Ninth Congress, First Session on H.R. 3170: A Bill to Provide Aid for the Development of Public Airports and to Amend Existing Laws Relating to Air Navigation Facilities, May 15, 16, 22, 23, 24, 25 29, 30, 31 and June 1 and 5, 1945 (Washington D.C.: Government Printing Office, 1945), 44–47, 66–69, 156–73, 262–73, 326–29, 385–97; Donoh W. Hanks, Jr., "Neglected Cities Turn to U.S.," *National Municipal Review* 35 (Apr., 1946): 172–76. This was also part of a larger conflict between the federal government and states over the regulation of aviation. See William C. Green, "The War against the States in Aviation," *Virginia Law Review* 31 (Sep. 1945): 835–864.

39. John R. M Wilson, *Turbulence Aloft: The Civil Aeronautics Administration Amid Wars and Rumors of Wars, 1938–1953* (Washington, D.C.: U.S. Department of Transportation, Federal Aviation Administration, 1979), 175–84; Advisory Committee on Intergovernmental Relations, *State Involvement with Federal Local Grant Programs: A Case Study of the "Buying In" Approach* (Washington, D.C.: Advisory Committee on Intergovernmental Relations, December 1970), 23. For a contemporary evaluation of the historic role of states in airport development, see *A Staff Report on Federal Air to Airports* (submitted to the Commissions on Intergovernmental Relations, June 1955), 86–92.
40. The National Airport Plan of 1944 defined the classes (designating them Class 1–5 or Class I–V) of airports as follows: "Generally, class 1 fields are suitable for private owner small type aircraft; class 2 for larger type private owner aircraft and small transport aircraft for local and feeder service; class 3 can accommodate present day twin-engine transport aircraft, while classes 4 and 5 serve the largest aircraft now in use and those planned for the immediate future." A Class I airport might not even have a runway, just a landing area of 1800 to 2700 feet long. A Class II airport had a runway 2700–3700 feet long; Class III, 3700–4700; Class IV, 4700–5700; and Class V, 5700 and over. Committee on Interstate and Foreign Commerce, *National Airport Plan: Letter from the Acting Secretary of Commerce Transmitting A Report of a Survey of the Need for a System of Airports and Landing Areas Throughout the United States, November 28, 1944* (Washington, D.C.: U.S. Government Printing Office, 1945), 4–5.
41. Allan C. Perkinson, "The Proposed National Airport Plan," *Virginia Law Review* 31 (Mar., 1945): 438–447; Wilson, pp. 171–174.

42. Wilson, *Turbulence Aloft*, pp. 164–168, 268. For more on the ideas of the winged gospel, and particularly the predictions concerning the widespread use of personal aircraft, see Joseph Corn, *The Winged Gospel: America's Romance with Aviation, 1900–1950* (Oxford and New York; Oxford University Press, 1983), 91–111.
43. "T. P. Wright Announces 1947 Airport Allotments," *Airports* (Feb. 1947): 20–21.
44. "Air Power Boosted by Increasing Airports," *Aviation Week* 52 (February 27, 1950): 63–64; "Where Federal Aid Cash Will Go," *Aviation Week*. 53 (November 13, 1950): 55.
45. Wilson, 184, 186–187, 191. For additional information on the early federal airport aid program, see Paul Barrett, "Cities and Their Airports: Policy Formation, 1926–1952," *Journal of Urban History* 14 (Nov., 1987), 112–137.
46. Katherine Johnsen, "House Opens Fire on Airport Aid," *Aviation Week* 58 (April 20, 1953): 14; "Congress Fights Airport Cuts," *Aviation Week* 58 (May 4, 1953): 16–17; "Committee Boosts Airport Aid Funds," *Aviation Week* 58 (June 8, 1953): 19–20; "Civil Aviation Faces Budget Slash in '55," *Aviation Week* 59 (December 14, 1953): 14.
47. "Airport Aid," *Aviation Week* 60 (April 26, 1954); 18; "Airport Funds," *Aviation Week* 60 (May 17, 1954): 16; Wilson, p. 22.
48. "Airport Funds," 16; "Senate, House Groups Oppose Airport Bill," *Aviation Week* 61 (July 12, 1954): 24, 26; Wilson, 38–42.
49. Wilson, 89–90; "$63 Million Annual Airport Aid Wins Unanimous Senate Approval," *Aviation Week* 63 (July 4, 1955): 83; "Record Airport Aid Bill Permits Funds for Terminal Construction," *Aviation Week* 63 (July 25, 1955): 116; "CAA Liberalizes Airport Aid Policy," *Aviation Week* 63 (October 17, 1955): 14; "Airport-Less Towns in Line for U.S. Aid Under New CAA Policy," *Business Week*, (October 22, 1955): 57.
50. The FAA remained an independent agency until 1967. President Lyndon Johnson, concerned by what he saw as the federal government's uncoordinated actions in the area of transportation, asked Congress for authorization to create a new cabinet-level department that would oversee all major federal transportation responsibilities. Upon the creation of the DOT on April 1, 1967, the FAA became one of many organizations under the new department and was renamed the Federal Aviation Administration.

51. "Hearings Begin on Bill To Boost Airport Aid," *Aviation Week* 70 (January 26, 1959): 38; Katherine Johnsen, "Democrats Push Airport Aid Measure," *Aviation Week* 70 (February 2, 1959): 45–46; Wilson, 231–233.
52. Johnsen, "Democrats Push Airport Aid Measure," 46.
53. Wilson, 233–234; "A Bad Case of Vetophobia," *The Nation* 188 (June 27, 1959): 565–566. Federal aid for airports was not the only form of federal support to local governments that came under budget scrutiny during the 1950s. For a discussion of the Eisenhower administration's efforts to reduce federal spending through reducing or eliminating a number of federal aid programs, including those for HUD, see Roger Biles, *The Fate of Cities: Urban America and the Federal Government, 1945–2000* (Lawrence: The University Press of Kansas, 2011), 46–81.
54. "Millions Voted for Airport Aid," *The Pittsburgh Press*, October 4, 1966; Kent, 191–197; Address, "As prepared for Delivery by David D. Thomas, Deputy Administrator, Federal Aviation Agency, Airport Operators Council, International, Phoenix, Arizona, October 14, 1966," FAA Information, Office of Information Services, Federal Aviation, Agency, Washington, D.C., RG 237, Box 207, File: National Airport Plan; "Airport Aid Allocations Cut $29 Million," *Aviation Week & Space Technology* 85 (December 26, 1966): 30; Karl M. Ruppenthal, "No Money for Safe Landings," *The Nation* 207 (August 26, 1968): 142–144.
55. Kent, 212–213; "Congress Unit to Urge More Aid for Development of Civil Airports," *Aviation Week & Space Technology* 86 (January 2, 1967): 28.
56. Kent, 213.
57. The AOCI began as the Airport Operators Council (AOC) in 1948. The AOC became the Airport Operators Council International (AOCI) in 1966. In the early 1990s, it changed its name once again to the Airports Council International-North America.
58. Ibid., 213–215.
59. Ibid., 216–217.
60. Ibid., 218–219.
61. "Senator Expresses Doubts About Federal Airport Aid," *The Miami News*, August 17, 1967.
62. Kent, 219; "10-Year Aviation Facilities Plan Readied," *Aviation Week & Space Technology* 88 (January 29, 1968): 32–33.

63. "10-Year Aviation Facilities Plan Readied," 32–33.
64. Quoted in "Airport, Airways Fund Plans Hit, *Aviation Week & Space Technology* 88 (May 27, 1968): 38.
65. "Airport, Airways Fund Plans Hit," 37–38; Donald C. Winston, "Airway, Airport Formula Revised," *Aviation Week & Space Technology* 88 (June 24, 1968): 39–40; Kent, 226–227.
66. Kent, 244–245.
67. Ibid., 245–251.
68. See Eric Henry, *Excise Taxes and the Airport and Airway Trust Fund, 1970–2002* (Washington, D.C.: Statistics of Income Bulletin, 2003), 44.
69. See FAA Historical Chronology, 1926–1996, "September 2, 1975" http://www.faa.gov/about/media/b-chron.pdf (accessed June 19, 2012).
70. For a discussion of the Nixon Administration's New Federalism and the push for block grant programs, see Biles, *The Fate of Cities*, 160–199, 189–190.
71. Donald C. Winston, "Nixon Transportation Trust Fund Idea Hit," *Aviation Week & Space Technology* 94 (March 29, 1971): 27; Charles E. Schneider, "Volpe Pledges Aviation Tax Fund Integrity," *Aviation Week & Space Technology* 94 (May 3, 1971): 23.
72. Donald C. Winston, "Senate Would Halt Trust Fund Loss," *Aviation Week & Space Technology* 95 (July 26, 1971): 14–15; _____., "Congressional Conferees Restrict Aviation Trust Fund Expenditures," *Aviation Week & Space Technology* 95 (August 2, 1971): 16–17; "Senate Sponsor Sees Nixon Veto of Trust Fund Use Restriction," *Aviation Week & Space Technology* 95 (October 18, 1971): 26.
73. "FAA Hit by House Unit for Lag in Airports/Airways Obligations," *Aviation Week & Space Technology* 96 (May 29, 1972): 21; "Airport, Airways Cost Scrutinized," *Aviation Week & Space Technology* 98 (January 15, 1973): 57, 59.
74. "Trust Fund Raid Effort Has Little Hope," *Aviation Week & Space Technology* 102 (April 21, 1975): 33–34.
75. William A Shumann, "House Unit Would Boost Airport, Airway Funding," *Aviation Week & Space Technology* 103 (September 29, 1975): 25; "Accord Reached by Conferees on Bill for Airport, Airway Aid," *Aviation Week & Space Technology* 104 (June 7, 1976): 34; "Airport Aid Funding Resumed After Year," *Aviation Week & Space Technology* 105 (August 30, 1976): 30.

76. Preston, 238.
77. For an example of the charge that the surplus in the aviation trust fund was being used to mask the size of the federal deficit, see "Air safety's unspent billions," *Business Week* (February 20, 1978): 55, 58, 60. That argument continued into the 1980s and 1990s. For example see Robert D. Hershey, Jr., "Washington Talk: The Aviation Trust Fund; A Fight Builds Over Money Not Spent," *New York Times*, September 4, 1987.
78. For more on the strike and its ramifications for the nation's air traffic system, see McCartin, *Collision Course: Ronald Reagan, the Air Traffic Controllers, and the Strike that Changed America*.
79. "Bill Would Alter Airport Funding Method," *Aviation Week and Space Technology* 111 (September 3, 1979): 40.
80. "Grants to Largest Airports Not Necessary, Bond Says," *Aviation Week & Space Technology* 111 (September 17, 1979): 28; "Editorial: Trusting the Trust Fund," *Aviation Week & Space Technology* 111 (October 22, 1979): 9.
81. "Editorial: Airports and the Trust Fund," *Aviation Week and Space Technology* 111 (November 19, 1979): 11; "Airport Operators Seeking Cannon Bill Modifications," *Aviation Week & Space Technology* 111 (December 3, 1979): 36.
82. "Cannon Willing to Discuss Head Tax in Airport Bill," *Aviation Week & Space Technology* 112 (January 21, 1980): 36; "The U.S. Skimps on Airport Safety," *Newsweek* XCV (March 31, 1980): 60; "Trust Fund Meets House Opposition," *Aviation Week & Space Technology* 112 (April 21, 1980): 31; "Airport Fund Delay Draws Opposition," *Aviation Week & Space Technology* 113 (August 18, 1980): 36; Michael Feazel, "Airport Aid Delay Until 1981 Expected," *Aviation Week & Space Technology* 113 (October 13, 1980): 36.
83. "Bill Would Drop Grants to 69 Airports," *Aviation Week & Space Technology* 114 (March 23, 1981): 34; "New Administration Backs Airport Defederalization," *Aviation Week & Space Technology* 114 (April 6, 1981): 33; "Opposition Arises to Airport Plans, Fees," *Aviation Week & Space Technology* 114 (April 6, 1981): 33; "House, Senate Airport Aid Bills Differ," *Aviation Week & Space Technology* 114 (May 4, 1981): 29; "A collision course on financing airports," *Business Week* (May 18, 1981): 46–47.
84. "Trust Fund Issues Cloud Airport Air Legislation," *Aviation Week & Space Technology* 116 (March 22, 1982): 36. In March 1982, the

ATA, a group presenting the nation's airlines, proposed a compromise that involved the federal government dropping plans to defederalize the larger airports. The compromise actually had the support of Senator Cannon who agreed to drop the defederalization plan if it meant passage of a multi-year airport aid program. The ATA also thought it had the support of the AOCI. Instead, the following month, the AOCI called for a voluntary defederalization program, splitting the aviation community on the airport aid measure once again. "Two Airport Groups Shift Defederalization Position," *Aviation Week & Space Technology* 116 (April 5, 1982): 38.

85. See Tax Equity and Fiscal Responsibility Act (P.L. 97–248), Title V: Airport and Airways Improvement Act of 1982; Surface Transportation Assistance Act of 1982 (P.L. 97–249); Airport and Airway Safety and Capacity Expansion Act of 1987 (P.L. 100–223).
86. James Ott, "Plans to Fund New Airports Stir Battle in Congress," *Aviation Week & Space Technology* 132 (February 12, 1990): 132.
87. See FAA Historical Chronology, 1926–1996, "November 5, 1990" and "May 22, 1991" http://www.faa.gov/about/media/b-chron.pdf (accessed June 19, 2012).
88. "Congress Weighs Airline Taxes," *Aviation Week & Space Technology* 146 (February 17, 1997): 48; James Ott, "FAA Funding Crisis Bolsters Ticket Tax," *Aviation Week & Space Technology* 146 (February 10, 1997): 28, 30.
89. Edward H. Phillips, "FAA Reform Bill Faces Fight From General Aviation," *Aviation Week & Space Technology* 143 (September 18, 1995): 31–32; _____. "FAA User Fees Cloud Reform Initiative," *Aviation Week & Space Technology* 143 (October 2, 1995): 34.
90. Edward H. Phillips, "Airports, Air Services Targeted for Cuts," *Aviation Week & Space Technology* 142 (February 6, 1995): 43.
91. Edward H. Phillips, "Congress Clears Path for FAA Reform," *Aviation Week & Space Technology* 143 (October 30, 1995): 40; Carole A. Shifrin, "FAA Reforms Set in Motion," *Aviation Week & Space Technology* 144 (April 1, 1996): 30.
92. Henry, *Excise Taxes and the Airport and Airway Trust Fund*, 44–47.
93. James T. McKenna, "Hill Lobbyists See No Quick Fix on FAA Funding," *Aviation Week & Space Technology* 150 (January 25, 1999): 49–51; _____. "FAA Would Tap Trust Fund to Finance Operations in 2000," *Aviation Week & Space Technology* 150 (February 8, 1999): 32; Paul Mann, "FAA's Budget Fate Remains Uncertain," *Aviation*

Week & Space Technology 151 (October 11, 1999): 29–31; James Ott, "AIR21 Fails in Congress," *Aviation Week & Space Technology* 151 (November 29, 1999): 41; Paul Mann, "FAA Budget Breakthrough Reached Just in Time," *Aviation Week & Space Technology* 152 (March 6, 2000): 36–37.

94. Henry, pp. 47–48; See also Robert S. Kirk, "Airport Improvement Program Reauthorization Legislation in the 106th Congress," Congressional Research Service, The Library of Congress, April 17, 2000.

PART II

Response to the Jet Age

Chapter Three: Response to the Jet Age: Federal–Local Interaction and the Shaping of the Aviation Landscape

The introduction of jet aircraft along with the rapid growth within the airline industry required expansion and modernization of the nation's airport infrastructure. Both local governments and the federal government played important roles. As the owners and managers of the airports, local governments struggled to keep up in an environment of often rapid and repeated change. Local officials often faced deciding whether to expand their existing airports or build new. As for the federal government, it provided important funding but also periodically turned surplus military bases over to local authorities. Some cities transformed these former military bases into civilian airports. In such cases, though, even during decades when building entirely new major airports proved extraordinarily difficult—so much so that very few completely new major commercial airports were constructed—local governments proved very deliberate and selective in choosing whether or not to take advantage of the former federal facilities' potential as commercial airports. Therefore, in determining exactly where and when new major airports would be established, the shaping of the aviation infrastructure remained a product of negotiation between federal and local officials, with local officials generally having the final word.

J.R. Bednarek, *Airports, Cities, and the Jet Age*,
DOI 10.1007/978-3-319-31195-1_4

If at First You Do Not Succeed...: Federal Efforts to Promote Regional Airports/Jetports

Perhaps one of the best examples of the influence of local voices can be seen in examples of repeated efforts of the CAB and the CAA/FAA to promote the construction of large regional airports or jetports to serve metropolitan regions, an idea promoted immediately after World War II (WWII) and again in the 1950s and 1960s in response to the introduction of jet airliners.[1] In case after case—both in terms of the failed efforts and the successful effort in the Dallas-Fort Worth area—local factors rather than federal action determined the outcome.

As noted, local airport officials scrambled to meet the requirements of the new jet airliners in the late 1950s and early 1960s. How they coped varied. Many decided to expand existing airports, while others looked for additional options. In the case of Chicago, airline traffic shifted from older, smaller Midway to the new O'Hare Airport, developed on a former military field the city acquired in 1945. Many in aviation—both local and federal officials—believed that the jet age required entirely new airports—airports designed and built specifically for jet airliners. When speaking of these new types of airports, proponents often used the term "jetport" to distinguish the new type of facility from older airports. And these "jetports" were also envisioned as large facilities that could serve metropolitan regions, rather than individual cities. Whether called regional airports or jetports, from the 1940s through the early 1970s, and especially after the advent of the jet age, federal aviation officials encouraged—and in some cases pushed—cities to build new, larger aviation facilities.

Three times over the course of three decades, proposals emerged for a regional airport or, later, a jetport to serve the Omaha, Nebraska, metropolitan area. The first proposal involved a joint Omaha-Council Bluffs, IA, airport and the last two a joint Omaha-Lincoln, NE, airport. Each of the proposals arose at times when the boosters or leadership of Omaha's airport were engaged in planning for expanded air travel. In each case, the proposal briefly garnered some attention and then quietly disappeared. In the 1940s and again in the 1960s, the proposals had a certain level of federal visibility and involvement. But in all cases, local opposition, or perhaps disinterest, brought an end to these discussions.

Even as WWII continued, Omaha city leaders looked to the future and began shaping an ambitious plan for civic improvements. As early as 1943, a number of local committees, under an umbrella mayor's advisory committee, had convened to explore various civic improvements, including expansion of the municipal airport. In early 1944, the *Omaha World Herald* reported that the airport committee sought a meeting with civic leaders in neighboring Council Bluffs, Iowa. The purpose of the meeting was to explore possible inter-state cooperation on airport improvements "beneficial to both cities." In explaining the context for the meeting, the chairman of the committee, Walter S. Byrne, referenced other inter-state airport development activity. He declared that a number of towns in Iowa already were exploring developing airports jointly with cities in either Nebraska to the west or Illinois to the east. He also pointed out that the twin cities of Minneapolis and St. Paul had recently initiated a cooperative airport effort and, despite the fact that both those cities were in the same state, he saw the issues there as similar to those that would be faced by Omaha and Council Bluffs.[2] Omaha and Council Bluffs are, to an extent, twin cities, bordering each other on the Missouri River.

Byrne's argument for a joint airport did reflect contemporary trends as for at least at some of the nation's smaller commercial airports joint ownership and management was becoming more common.[3] However, despite the existence of a number of potential examples of joint airport projects, little came of the initial discussions in the Omaha area. In fact, the next reference to Council Bluffs in connection to airport planning apparently came not from local planners but from federal officials. In late 1944, the CAA, preparing the post-war airport plan in anticipation of congressional action on the airport aid bill, called for the construction of a so-called "super airport" for the Omaha area. Omaha's airport boosters favored a site west of the city. However, a few days after the first mention of the "super airport," CAA officials in Washington informed local airport planners that the CAA anticipated that the new airport would be built on a site in Council Bluffs. J.B. Bayard, Jr., the chief of the CAA's planning and survey division, said that Omaha had failed to develop a "comprehensive aviation plan." Therefore, CAA specialists had to come to the area to determine the best location for any new large airport and had tentatively settled on a site on the Iowa side of the Missouri River.[4]

In early 1945, Omaha and Council Bluffs aviation boosters met to discuss this proposed "super airport," but failed to come to any agreement. Not surprisingly, the Council Bluffs boosters steadfastly supported the Iowa site, even though apparently the CAA later denied the site had been selected, while Omaha boosters held out for a site in Omaha.[5] The primary outcome of the controversy proved an incentive to Omaha's planners to finalize their proposals for the existing municipal airport. Those plans formed the basis of the airport section of the so-called "Blue Book" plan, the collection of major civic improvements developed out of the planning efforts begun in 1943. The airport section called for construction or extension of a number of runways, a doubling of the size of the administration building, and the establishment of at least one satellite airport. In 1946, local voters approved both general obligation bonds to fund improvements and the creation of a special airport commission to oversee the spending of the bond money.[6]

The second campaign for a regional airport arose as part of another ambitious civic improvement plan. Like the "Blue Book" before it, the "Omaha Plan" included a number of separate proposals for bond issues to finance a whole host of civic improvements, including expansion of the airport to prepare for the jet age.[7] In the midst of the development of the "Omaha Plan," in June 1957, State Senator John P. Munnelly floated the idea of building a joint Omaha-Lincoln airport. Lincoln was the second largest city in Nebraska, the state capital, and located about 50 miles from Omaha. Officials in Omaha, however, continued to plan for the expansion of the local airport. The idea, though, apparently had some staying power. It came up again briefly in May 1959, following the defeat of the airport bond issue. Despite the bond issue failure, the newly created Omaha Airport Authority "politely" informed the Lincoln Airport Authority that it was not interested in a joint project.[8]

The third proposal for a joint airport came in 1969 and, like the proposal in the 1940s, involved federal officials. In June 1970, the *Omaha World-Herald* reported that, according to State Planning Director Douglas Bereuter, the next iteration of the FAA's national airport plan would "describe Omaha's Eppley Airfield as 'hopelessly obsolete' for jumbo jet traffic and [would] recommend a joint Omaha-Lincoln airport."[9] City and airport leaders reacted strongly—both positively and negatively—to the report. Omaha city council member Al Veys supported the idea of joint airport, predicting the need for an international airport located between Omaha and Lincoln within the next 25 years. State Senator John E. Knight of Lincoln, described as a long-time supporter of a joint airport, believed the airport "could be financed through a combination of federal, state, Lincoln and Omaha tax dollars." On the other hand, Omaha city council member Betty Abbott, a strong airport booster,[10] opposed the proposal. She believed that it made little sense given the expansion program currently underway at the airport. The executive director of the airport authority, Ronald B. Grear, and Chairman, James B. Moore, echoed her comments, declaring that Eppley Airfield would be sufficient to meet the needs of the Omaha area for another 25–30 years. Grear also pointed out that no local tax dollars were involved in any of the recent airport improvements. Instead, airport revenue bonds had financed them.[11]

Within days of the initial report, an FAA official responded to the local controversy it had generated. He indicated that the idea for the joint airport had come not from federal sources but from the Nebraska Department of Aeronautics, though the proposal would be mentioned in the update to the national airport plan for 1969. However, the official denied that the plan called the existing airport "hopelessly obsolete." Those were the words of the state planning director, not the FAA. Further, the FAA official said that the state plans actually called for the continued development of both the Omaha and the Lincoln airports, but indicated that the idea of a joint airport was suggested primarily to promote additional thinking about future aviation requirements.[12] Seemingly, support for a joint Omaha-Lincoln airport existed primarily at the state level, with only token support from the FAA. Local opposition scuttled the idea.

Local opposition, as well as disinterest, also played a role in scuttling a plan for a regional airport to serve southwestern Ohio, to be located between the cities of Cincinnati and Dayton, also approximately 50 miles distant from one another. Dayton had a municipally owned commercial airport located 11 miles north of the city near Vandalia, Ohio. Cincinnati, however, no longer owned and operated a major commercial airport. In 1937, a catastrophic flood of the Ohio River inundated Cincinnati's Lunken Airport. Following the flood, a group of aviation boosters in nearby Kenton County, Kentucky, announced plans for a new airport, to be built in Boone, County, Kentucky. (Kentucky's airport enabling law allowed one county to own and operate an airport in another county.)[13] While some preliminary work was done, extensive action awaited the 1940s. With the outbreak of WWII, both the Kenton County Airport Board and the city of Cincinnati sought federal aid to build military airfields. The CAA decided to fund construction at the Boone County airport site and rejected the application from Cincinnati. As early as 1943, airlines serving the Cincinnati area reported that they planned to move their operations from the small, flood-prone Lunken Airport to the new Kentucky facility after the war In June 1944, Kenton County Airport Board officials appeared at a hearing of the Cincinnati city council. At that meeting, they proposed renaming the facility "The Greater Cincinnati Airport" and noted that it "was planned and financed with a view to affording Cincinnati a great air terminal for the use of the Greater Cincinnati area." As announced, the commercial airlines serving the Cincinnati area moved their operations out of Lunken and to the new Greater Cincinnati Airport in January 1947.[14]

The idea for an Ohio regional jetport to serve Cincinnati and Dayton arose in early 1961. A group headed by Dayton's mayor, R. William Patterson, publicized the results of a consulting engineering firm's report that concluded such a facility would be needed as early as 1965. Without such a facility, the report warned, the region's potential for economic development and growth would be harmed. The FAA response to the proposal was mild, but favorable. A spokesperson said that the FAA was generally in favor of such proposals, but could not comment further until it received a formal request. Locally, many Dayton officials and the local newspaper supported the program, while Cincinnati city officials and its local newspapers reacted initially with some concern, eventually deciding against the proposal.[15]

Despite the opposition from the City of Cincinnati, throughout the rest of 1961, proponents worked to drum up support in an 11-county area. They envisioned the creation of a regional airport authority to plan, construct and then manage the new facility. Hamilton County (location of Cincinnati) officials tentatively came out in support of the creation of the regional authority. The manager of the Greater Cincinnati Airport, Arven H. Saunders, however, began appearing at some of the meetings called to discuss the proposal. He strongly objected to the idea, arguing that the Greater Cincinnati Airport could serve the needs of Cincinnati far better than any regional airport built between that city and Dayton. His airport was closer, only 12 miles from downtown Cincinnati, while the new airport would be at a greater distance.[16] And the Kenton County Airport Board was involved in an expansion program to meet needs well into the future.[17] Opinions in Dayton also became divided. Though the mayor of Dayton was leading the campaign, the city manager, Herbert W. Starick, opposed it, fearing that it could jeopardize federal aid for Dayton's own municipal airport. The Dayton Area Chamber of Commerce echoed the city manager's concerns. The local newspaper, however, sided with the mayor.[18]

The FAA was ambiguous in its comments. FAA Administrator Najeeb Halaby, addressing questions from Cincinnati, stated that "if Cincinnati wants a regional airport, the community, and not the federal government, will have to provide the initial push and plans for it." He stated that he was neither opposed to nor supportive of the idea, calling it a local issue. On the other hand, he said that if a workable plan was forwarded to the FAA "it would probably get help from federal funds to build it."[19]

The legislature of the state of Ohio did show support for the effort, passing the enabling law needed for the creation of the regional airport authority. The success of the new organization, however, depended upon having each county in the 11-county region agree to participate. Following the passage of the legislation, a few counties appointed representatives, but after the first two counties (Clinton and Butler) agreed to participate, the minimum needed under the legislation to establish the new authority, most counties decided against involvement, with only Montgomery County (location of Dayton) voting in the affirmative. Most significantly, Hamilton County commissioners voted against participation in September 1962. By the time of the Hamilton County vote, it was also clear that the major airlines serving Cincinnati and Dayton also opposed the regional airport idea.[20] Though apparently there was an attempt to continue without the participation of Hamilton County, the lack of support in the Cincinnati area effectively killed the proposal.[21]

The FAA's rather passive response to the regional airport in the Dayton-Cincinnati area stands in stark contrast to its more aggressive stance in Dallas and Fort Worth, where pressure from it and the CAB did result in the construction of a new regional jetport. The events that led to the construction of the Dallas-Fort Worth International Airport in the late 1960s actually began in the late 1930s. Both Dallas and Fort Worth—approximately 35 miles distance from one another—had small commercial airports. Following the passage of the Civil Aeronautics Act in 1938, the CAA, as part of the first national airport plan, approached the two cities with a proposal to build a single regional airport. When the two cities balked at the idea, the government turned to a nearby small town, Alliance. Two Texas-based airlines, American and Braniff, agreed to buy land and deed it to the town. The CAA would then build the runways. This temporarily encouraged both Dallas and Fort Worth to re-enter negotiations with the CAA and a tri-city agreement was established. However, as planning continued, Dallas officials objected to the master plan, which placed the air terminal at a location that favored Fort Worth. That, plus personality clashes between the mayors of the two cities, resulted in Dallas withdrawing from the pact. Instead, the government constructed a military field at the site during WWII. After the war, the government declared the field surplus and Alliance agreed to deed the land to Fort Worth. With encouragement from the CAA, Fort Worth decided to develop it as its post-war airport.[22]

Dallas also profited from wartime spending on municipal airports as the U.S. Army Air Corps expended $6 million to improve Love Field as a major base of operations for Ferrying Command. Though discussions over development of a joint airport resumed after the war, Dallas city officials, following the advice of outside consultants, decided instead to further improve and expand Love Field. This then set off a prolonged battle between the two cities for regional air supremacy. Both Fort Worth and Dallas petitioned the CAB for additional air service. Despite renewed pressure from Washington, throughout the 1950s Dallas leaders refused to accept the notion of transforming Fort Worth's post-war airport, Amon Carter Field (later Greater Southwest International Airport), into a joint regional airport. The urban rivalry between the two cities was intense as each viewed local air service as vital to economic development.[23]

Federal pressure continued from both the FAA and CAB into the 1960s. Both agencies were convinced that Dallas and Fort Worth—each having airports with limited capacity for expansion and improvement to serve jet airliners—would be best served by a single regional jetport. The issue finally came to a head in 1964 when the CAB gave the cities 180 days to agree on a location for a new regional airport or it would chose a location for them—with the implication that jet traffic would not be allowed at either Love Field or Greater Southwest International. Though the process took longer than the original 180 days, in May 1965, the city councils of both communities agreed to a plan for a single regional airport. The designated airport site was just north of Fort Worth's Greater Southwest International Airport (soon to be abandoned) and included at least 10,000 additional acres, not including clear zones, which raised the total airport area to 21,000 acres.[24]

While pressure from the federal government was important, historian Robert Fairbanks argued that the two cities finally agreed to the single regional airport due in large part to local factors including the fact that growth and development in the two cities had essentially resulted in the creation of a single metropolitan region. Whereas in the 1940s and 1950s, Dallas and Fort Worth viewed themselves at the centers of separate metropolitan areas, by the mid-1960s, leaders in both cities, but particularly in Dallas, had developed a vision of a single region.[25] Construction on the new airport began in 1969 and Dallas-Fort Worth International Airport opened in 1973.

The Everglades Jetport: The Most Famous Airport That Almost Never Was

Just as construction was set to begin on the Dallas-Fort Worth airport, the federal government was involved in another controversial jetport proposal. In July 1968, Dade County, Florida, announced plans to construct a new airport facility on a 34 square-mile site 40 miles from downtown Miami and 6 miles north of the Everglades National Park. As initially envisioned, the new airport would absorb the airline training flights conducted at Miami International Airport. It was hoped that moving the jet training flights to the new facility would reduce congestion and noise complaints at the Miami airport. At the time, airline training flights made up a little more than a third of the operations at Miami International Airport. While the new airport would at first only handle training flights, it was anticipated that in the future it would evolve into a new regional commercial jetport, one that would also allow for the use of planned supersonic commercial jet airliners.[26]

By early 1969, however, local conservationists sounded the alarm. They pointed out the dangers the new facility posed to the wildlife in the area, as well as the potential for water pollution from fuel spills. Concerns also focused on likely damage to the nearby Everglades National Park, the nation's only subtropical park. The park's superintendent, John C. Raftery, warned that development of the airport threatened not just wildlife and water quality, but also the quiet and solitude experienced by park visitors. Dade County officials also projected a great deal of commercial and residential development around the new facility. Raftery worried that such development would further threaten the water supply crucial to the survival of the park.[27]

By the summer of 1969, the controversy over the jetport—already under construction—extended to Washington and a struggle over the further development of the site erupted between the DOT and the Department of the Interior. In June 1969, at a Senate Interior Committee hearing on the jetport, Department of Interior lawyers issued an opinion stating that full development of the facility would threaten Everglades National Park and that the DOT had failed to comply with federal laws protecting the nation's natural resources. These hearings also came at about the same time Secretary of Transportation John Volpe and Interior Secretary Walter J. Hickel, appointed a joint committee to study the jetport's potential to create environmental problems.[28]

Meanwhile, local and national environmental groups continued to protest the jetport, while the director of the Dade County Port Authority, Alan C. Stewart, denied that the facility would create any environmental problems and called the conservationists "butterfly chasers." Adding to the controversy, the airport site included land used by the Miccosukee Indian tribe as hunting grounds and for an annual ritual known as the Green Corn Dance. The port authority stated that the development would preserve the huts, located along the Tamiami Trail, used for the dance. The tribe originally supported the project, viewing it as bringing needed jobs, but by summer 1969 tribal leaders joined theirs to the voices opposing the development, fearing damage to their hunting grounds as well as to the national park.[29]

In August 1969, the joint committee issued a report urging that an alternative site be found for any new jetport in the Miami region. It further recommended that the training facility then under construction also be moved. The report came about a month before training operations were scheduled to begin at the single runway facility already nearing completion. Though the report remained unpublished, in early September the Republican governor of Florida, Claude R. Kirk, Jr., added his voice to those opposed to the further development of the new facility. After a meeting between Kirk, Volpe, and Hickel in Washington, D.C., the three announced a tentative agreement. The one-runway training facility would open and operate at the site, but an alternative site would be found for any new major regional jetport. Though conservationists continued to protest any airport operations near Everglades National Park, the agreement was formally announced in early January 1970.[30] A *New York Times* article on the agreement included a statement from President Richard Nixon. While he apparently supported the outcome, he also suggested that federal government still had a role to play in developing needed new airport facilities. However, it was clear that strong local opposition—helped by the emerging national environmental movement—could force an end to federal support for new airport facilities.[31]

Government Surplus: Military Air Bases and the Civilian Aviation Infrastructure

During WWII, the federal government spent millions upgrading municipal airports to serve military needs. Following the war, the government returned those now enhanced air fields back to the cities. Therefore, a number of cities, both large and small, started the post-war period with vastly improved local airports due to their military use during the war. Those cities included Los Angeles, California; Atlanta, Georgia; Milwaukee, Wisconsin; and Dayton, Ohio. In addition, the government built new air bases across the country. Many cities, rather than continuing to use smaller, pre-war airports, elected to use surplus military airfields as their new post-war airports. Cities in that category included Chicago, Illinois, which, as noted, largely replaced its original municipal airport, renamed Midway after the war, with what became O'Hare International Airport; Pittsburgh, Pennsylvania; Spokane, Washington; Raleigh-Durham, North Carolina; and Midland, Texas. Many cities, thus, were able to meet the immediate post-war increase in demand for air travel due to the groundwork laid during WWII.[32]

After 1945, the government periodically declared other former military facilities as surplus—particularly during two waves of base closures first in the late 1960s and early 1970s and then again in the 1990s. Some cities did take advantage of the surplus fields. In a number of other cases though, the same objections that made new airport construction extraordinarily difficult—especially noise and other environmental impacts—fueled local debates and limited the number of former military facilities transformed into commercial airports in the post-war period.

One such case where local factors slowed and limited the development happened in the New York City metropolitan area. By the early 1970s, the PNYNJ had spent nearly a decade searching for a site for a fourth major jetport for New York City and its region. Over the years, the organization had explored potential sites in Morris County, New Jersey (the Great Swamp), Solberg, New Jersey, and even examined the idea of building an airport on landfill off Staten Island. All of the potential sites evoked strong negative reactions from both local interests and the airlines.[33] A round of Vietnam-era base closures provided another option. The Department of Defense announced the closure of Stewart Air Force Base, located more than 60 miles north of New York City, in 1970. Though Governor Nelson Rockefeller strongly championed it as a potential location for the long-sought fourth jetport, both the PNYNJ and the Metropolitan Transportation Authority—the state agency which had taken over planning for Stewart—rejected the site for various reasons. Nonetheless, in 1971, Governor Rockefeller persuaded the state legislature to appropriate $71 million to turn Stewart into a commercial airport.[34]

Immediately, however, the plan inspired very mixed feelings within the local community. On the one hand, the area around Stewart had been experiencing deep economic problems and the airport plan promised new jobs and increased economic development. And though many local politicians welcomed the potential economic benefits, others feared that it would bring chaotic growth and noise pollution. In addition, there had been local support for the development of Stewart as a general aviation airport. The president of the Chamber of Commerce, for example, supported such a facility, but strongly objected to the transformation of Stewart into a major jetport and promised that "we have absolutely no intention of letting them get away with it – wherever jetports have been resisted in recent years they have been defeated."[35]

Despite local objections, state officials moved forward with improvement plans, but efforts to actually entice airlines proved far more problematic. Campaigns to attract scheduled commercial airline service to Stewart began in the mid-1970s, but failed. In the mid-1980s, state officials launched another determined push. Though initially unsuccessful, the renewed effort came at a time when the local response to the commercial development of the field had changed from negative to positive as, since the early 1970s, several major businesses had located facilities at or near the airport. Success in attracting airline service finally came in the early 1990s and by the mid-1990s more than 400,000 commercial passengers used the airport. Compared to what was often described as the "lavish dreams" and "grandiose" plans of the 1960s, even local advocates admitted that Stewart Airport was best described as "just a nice country airport" and they were quite happy to keep it that way.[36] Well into the twenty-first century, though, efforts continued to transform the "nice country airport," managed by the PNYNJ since 2007, into the fourth major airport for New York City.

The same round of base closures in the late 1960s and early 1970s that provided the New York region with the Stewart site also presented the Los Angeles region with a potential location for a new airport near the small desert community of Palmdale. Efforts to transform the military facility (Air Force Plant 42) and the surrounding desert area into a major commercial airport began in the late 1960s. With support from the City of Palmdale and the USAF, the Los Angeles Department of Airports began purchasing land. Almost immediately, however, the effort ran into opposition from environmentalists. A California land task force headed by consumer activist Ralph Nader declared that an airport would be very harmful to what it termed "the deceptively fragile desert environment."[37] In addition, the Sierra Club and a group of local landowners filed suit. This temporarily halted the flow of federal funds to the project, yet a passenger terminal opened on June 29, 1971, on leased land. Land acquisition began again in 1972. By 1978, an environmental impact study concluded that the Palmdale site was appropriate for the development of a major airport facility to meet the growing demand in the Los Angeles region. In response, the Los Angeles Department of Airports acquired most of the land perceived as necessary for the project by 1983.[38]

In the meantime, however, Los Angeles had embarked on a massive airport improvement program for Los Angeles International Airport (LAX) in anticipation of the 1984 Olympics, hosted by the city. This included new international and domestic passenger terminals as well as upgrades to two east-west runways.[39] That was followed in 1989 with a second round of airport master planning. Once again, the issue of using the Palmdale site as a major airport facility was raised, this time by opponents of further expansion of LAX. Though local officials in Palmdale continued to avidly support airport development, opposition to the expansion of LAX rather than development of Palmdale stood at the heart of the political battle over the desert airport.

Three major issues hampered city efforts to promote the expansion of LAX: the city's budget crisis in the 1990s; community opposition; and environmental issues. Community opposition proved the most important as City Council member Ruth Galanter, whose district was next to LAX, championed the anti-expansion, pro-Palmdale effort. She was upset with the airport master planning process, which she viewed as forcing the people of her district to bear most of the environmental consequences of expansion. She devised the strategy of proposing Palmdale development as an alternative to LAX expansion.[40]

As the controversy over the expansion of LAX continued into the 1990s, Galanter and her supporters emphasized the positive aspects of Palmdale development. First, there was plenty of land and, second, the city already owned the land. Third, the desert area was sparsely populated. On the other hand, those in favor of LAX expansion emphasized the limitations of the Palmdale site. First, the Palmdale site was at a remote desert location and only one highway connected it with Los Angeles. Second, it was estimated that in the absence of major infrastructure improvements—including a planned, highly expensive, high-speed rail system—by 2020, a peak-hour trip from Los Angeles to Palmdale could take up to three hours and perhaps longer. Third, the small population meant a much more limited local market for air travel, especially when compared with a new alternative location, the soon to be closed El Toro Marine Air Station in heavily populated Orange County. Finally, opponents noted the high altitude of the field and the heat experienced in the area as well as the fact it was near a military supersonic test corridor.[41]

Eventually, the city attempted to compromise with opponents by both scaling back expansion plans at LAX and agreeing to work with the city of Palmdale on long-term development of the airport. Nevertheless, when the master plan was finally issued in early 2001, opponents continued their delaying tactics in hopes of postponing any action past the end of the term of Mayor Richard Riordan, who backed the LAX expansion plan. Then, in 2001, two events happened that significantly influenced the final outcome. First, James Hahn, an opponent of airport expansion, was elected mayor; second, the September 11, 2001, terrorist attacks. In the wake of those two events, plans for expansion at LAX were drastically reduced, largely in light of an immediate 20 % drop in traffic following the attacks and the new emphasis placed instead on security and modernization. Mayor Hahn, though, also continued to support development of Palmdale. The Palmdale airport had intermittently enjoyed commercial air service between 1971 and the early twenty-first century. In 2006, Los Angeles World Airports (LAWA, the city agency in charge of the Los Angeles's airports[42]), in cooperation with the city of Palmdale and local boosters, launched a major effort to bring commercial jet service to the airport. United Airlines, with a grant from the US DOT's Small Communities Air Service Program, began service from Palmdale to San Francisco in June 2007.[43] Nonetheless, Palmdale was still far from realizing the ambitions of its major supporters. In 2009, United Airlines canceled its flights, and after years of failed attempts to draw sustained commercial service to the airport, LAWA surrendered the commercial certification for the Palmdale airport.[44]

The Palmdale site was not the only former military base in the LA region with potential for conversion to a civilian airport. In the early 1990s, the post-Cold War base closure and realignment process involved no fewer than four military installations in the Los Angeles area—March, Norton, and George Air Force Bases, and, as noted, El Toro Marine Air Station. March Air Force Base became an air reserve base. The cities of San Bernardino and Victorville have worked to transform Norton and George, respectively, into commercial airports, mostly focused on air cargo shipments rather than commercial passenger traffic. Of the four bases, El Toro, located in Orange County and scheduled for closure in 1999, emerged as the most promising site for a major reliever airport to LAX. The existing Orange County Airport (John Wayne) was relatively small (500 acres) and agreements on noise abatement severely limited any potential for expansion. Proponents argued development of the 4700-acre El Toro site, located in the middle of well-populated area that also served as an important business center, would bring welcome economic benefits to an area viewed as underserved by commercial air service. Plans to transform El Toro into a major airport, however, immediately sparked heated controversy with both sides turning to the ballot box and the courts. Basically, the fight pitted pro-airport northern Orange County against anti-airport southern Orange County. Both affluent as well as working-class and immigrant residents in the north viewed the airport as a means to transform the area into a mecca for high-technology industries and to bring other forms of economic development. In the south, conservative, affluent suburban voters joined with liberal environmentalists to fight what they viewed as a threat to the environment and, more importantly, their suburban way of life.[45]

Overtime, an economic boom in northern Orange County diminished the appeal of the economic arguments in favor of the airport. Further, opponents argued the airport threatened to exacerbate the growing problem of soaring housing costs. Additionally, supporters made a number of tactical errors. For example, county planning officials in 1996 announced plans for a much larger facility than had been publicly discussed in the past, angering both residents and environmentalists. Even after backing down from the larger plan, the county faced lawsuits from environmentalists who forced it to conduct a second environmental impact study. The second study took four years and gave the opposition time to organize further and refine its arguments.[46] Though many efforts, including one aimed at electing anti-airport candidates to the county Board of Supervisors, failed, the anti-airport side proved unrelenting in its criticism and declared the proposed El Toro airport "unwanted, unneeded, and unsafe.'"[47] Finally, in 2002, following the September 11, 2001, terrorist attacks, anti-airport organizations placed Measure W on the ballot. Passed in March 2002, Measure W rezoned the El Toro site as parkland and open space. Under the new zoning, it could not be used as an airport. The success of Measure W effectively killed the last efforts to transform El Toro into another major international airport in the Los Angeles region.[48]

There were, however, a number of cities that did take advantage of the most recent base closures, including Austin, Texas. Austin opened its first municipal airport in 1930. Robert Mueller Municipal Airport served as a convenient, close-in facility for the city for several decades. By the 1970s, however, development surrounded and hemmed in the airport and both businesses and residents began to make increased noise complaints. Further, a number of airplane accidents near the airport, most involving smaller airplanes, prompted those living under the facility's flight paths to voice serious safety concerns. Faced with these issues, the Austin city council moved in 1976 to explore the possible joint use of nearby Bergstrom Air Force Base. The initial idea was to gradually develop the new facility and over eight to ten years move flight operations from Mueller to the new airport. The sense of urgency was raised when Southwest Airlines, then still an intra-state Texas airline, began flights out of Austin. Early hopes for the joint-use facility were dashed, however, when the Department of Defense turned down the request in 1978.[49]

While the city continued to hold out hopes for the eventual use of Bergstrom, during the 1980s, strong sentiment built locally for closing Mueller and finding an immediate alternative. However, there were those who favored keeping the conveniently located municipal airport. Further, Southwest Airlines, by that time a post-deregulation expanding interstate carrier, hinted that it would oppose a move to a new facility. In this unsettled atmosphere, the city council commissioned a number of studies on the airport issue, including some aimed at identifying a new airport site. By the mid-1980s, a location 18 miles north of the city near Manor, Texas, emerged as a favorite of those supporting construction of a new facility. In 1985, the city held a non-binding referendum on whether or not to create an airport at the Manor site. It failed by a narrow margin, but, as it was non-binding, the city council continued to examine that possibility. Meanwhile, air traffic continued to grow at Mueller, adding to the problems of noise and congestion. Finally, in 1987, local sentiment apparently changed as a second referendum indicated that voters strongly favored both building a new airport and closing Mueller.[50]

On the heels of the 1987 referendum vote, the city then began the process of preparing to build a new airport at the Manor site. However, a number of factors, including a controversy over naming a project manager, slowed down the process. Frustrated residents—both from the area around Mueller who wanted that airport closed and around the Manor site who wanted any land purchases finalized—pushed for legislation at the state level. The Texas state legislature passed a law dealing with noise abatement and worded such that it essentially set deadlines for either closing

Mueller or soundproofing nearby residences and businesses. The law also set a deadline for beginning land purchases for the new airport. Just as the process got underway, in January 1990, the Department of Defense announced the possible closure of Bergstrom Air Force Base. Viewed by many as a better—read less expensive—option, the announcement once again created controversy over which direction the city should take. Land owners near Manor insisted that the city either buy their land or formally end the project. Some in the city insisted that instead of welcoming the closure as an opportunity, the city should wage a battle to fight the base closure due to the jobs that would be lost. Others questioned whether the Bergstrom site was the best for a new airport. All the while, air traffic at the old facility continued to expand, resulting in a number of construction projects at Mueller aimed at keeping up with the growing number of passengers.[51]

When the Department of Defense and President George H.W. Bush finalized the base closure list in 1992, Austin's city council decided to end the effort to build a new airport at Manor and to instead transform the former air force base into a new municipal airport for the growing city. The following year yet another referendum indicated that the voters in Austin favored developing Bergstrom and closing Mueller. By September 1993, the final military units left Bergstrom AFB. The city then moved very quickly to provide the facilities needed to transform the military airfield into a civilian airport. The plan was to open the airport to cargo by 1996 and to passengers by 1998. Construction began in March 1995 and the cargo facility opened on schedule in 1996. The last commercial flights landed at Mueller Airport on May 22, 1998. Overnight all operations moved from the old field to the new facility where the first flights departed on May 23, 1998.[52]

From offering surplus military airfields to using the government's power to regulate commercial air traffic, federal officials attempted to provide new locations for or to prompt the construction of new commercial airports. The role of the federal government, though, remained limited as the ultimate decision-making remained local.

Notes

1. For a sense of the FAA/CAB plans, see Robert H. Cook, "CAB Regional Airport Plan Faces Delay," *Aviation Week & Space Technology* 80 (April 20, 1964): 45; Memorandum, May 20, 1964, Subject:

Agency Policy with Respect to Regional Airport Authorities and Navaid Installation Where Unused Facilities Exit, From: Director, Office of Policy Development, To: Administrator, RG 237, Box 119, File: Airports and Other Landing Facilities 1964. The idea of regional airports was also supported by the Airports Operators Council, International, see attch: "A Resolution Urging the President and Congress of the United States to Establish an Interstate System of Regional Airports," ltr from Bennet H. Griffen, Howard, Needles, Tammen & Bergendoff, Consulting Engineers, to Mr. David O. Thomas, Deputy Administrator, FAA, October 18, 1966, RG 237, Box 207, File: National Airport Plan.

2. "Two-City Air Tieup Sought: Omaha Unit Hope for Bluffs Meeting," *Omaha World Herald*, February 7, 1944 (Omaha Public Library Clip File, hereafter OPL).
3. As early as 1942, *American City* reported that six states (California, Idaho, Kansas, Oregon, South Dakota, and Washington) had recently passed enabling legislation allowing two or more cities, or cities and counties, to jointly own and operate airports. In a 1944 work, aviation law expert Charles Rhyne reported that by that date, 32 states had such legislation—many laws predating the actions referenced in the *American City* article—and that joint airports were actually in operation in 20 states. A second *American City* article, also in 1942, reported that 42 joint airports existed and that 55 more were likely in the near future. Most of those airports involved cooperation between cities and counties, but seven were multi-city facilities. Most of the joint airports were concentrated in the nation's smaller cities. Only two cities (not named and not indicated whether city-county or city-city) with populations exceeding 100,000 had jointly owned and managed airports. Of the remaining, 18 were associated with cities with populations between 10,000 and 100,000 and 22 with cities with populations below 10,000. "Multi-City Airports Increase," *The American City* 57 (October 1942): 13;"Legislation Gave Cities Broader Airport Control," *The American City* 57 (April 1942): 11; Rhyne, *Airports and the Courts*, 39–42.
4. "Superport Neglected by Omaha: 'City Had No Plans,' so CAA Specialist Favored Bluffs Site," *Omaha World-Herald*, December 6, 1944 (OPL).
5. "Reply Held Up on Bluff Site: Airport Meet Cordial but Indecisive," *Omaha World-Herald*, January 4, 1945 (OPL).

6. The *Omaha World Herald* published a series of articles, usually on the front page, supporting the various parts of the so-called "Blue Book" plan. The article on the airport was the first and it appeared on March 23, 1946 (OPL). For more on the Blue Book plan see Janet R. Daly-Bednarek, *The Changing Image of the City: Planning for Downtown Omaha, 1945–1973* (Lincoln: University of Nebraska Press, 1991), 108–120. "Bond election results," *Omaha World Herald*, November 6, 1946 (OPL).
7. For information on the Omaha plan see Harl Dahlstrom, *A. V. Sorenson and the New Omaha* (Omaha, NE: Lamplighter Press, Douglas County Historical Society, 1987), 20–40, 690–71; Daly-Bednarek, *Changing Image of the City*, 124–143.
8. "Airport Plan is Two-City," *Omaha World-Herald*, June 9, 1957 (OPL); "Omaha-Lincoln Airport Proposal Turned Down," *Omaha World-Herald*, May 14, 1959 (OPL).
9. In 2014, the longest runways at Omaha's Eppley Airfield measured approximately 8000 feet. The first 747s called for runways nearly 11,000 feet long. Unless airport authorities could find the space to lengthen the runways by approximately 3000 feet, and the state aviation director had doubts that could be accomplished, the airport would be "hopelessly obsolete" for the new jumbo jets.
10. The road from downtown Omaha to the airport is named Abbot Drive.
11. Robert Hoig, "Grear: Eppley Not Obsolete in Jumbo Age," *Omaha World-Herald*, June 7, 1969 (OPL); "Joint Airport could Cost $100 Million," *Omaha World-Herald*, June 8, 1969 (OPL).
12. Darwin Olofson, "Joint Airport Idea a 'Spur to Thinking'," *Omaha World-Herald*, June 13, 1969 (OPL).
13. Arnold Knauth, et al., eds., *U.S. Aviation Reports, 1928* (Baltimore: United States Aviation Reports, Inc., 1928), 537–539.
14. "Detailed History" http://www.cvgairport.com/about/history2.html (accessed May 2, 2012); "U. S. Plans Airport for Covington: Aeronautics Bureau Submits Project," *Cincinnati Post*, September 28, 1940; "City Loses Out on New Airport; Ky. Site OK'd, $2,000,000 Allotted For Huge Field Near Covington," *Cincinnati Post*, October 1, 1942; "Kenton Field Seen as New Air Hub," *Cincinnati Post*, June 8, 1943; "Kenton Airport Offered to ATC, Ferrying Unit; Two Services May Use Field as Base; Army Confirms $100,000 Project at Lunken Cancelled," *Cincinnati Post*, March 4, 1944; "Use of Airport

to be Offered," *Cincinnati Post*, June 7, 1944; "Kenton Airport Nears Completion as Talk of New Field Here Fades," *Cincinnati Post*, August 24, 1944.

15. Dick Eben, "S.W. Ohio Called Hub of Future Air Traffic: Experts Say Need Will Be Acute by '65," *Dayton Daily News*, March 12, 1961; Jack Jones, "Airport Work Urged by FAA: Regional Plan Not Listed, But Agency Seen Favorable," *Dayton Daily News*, April 25, 1961; Bill Kagler, "Many Questions To Be Answered By Regional Airport Proponents," *Cincinnati Enquirer*, June 1, 1961.
16. The proposed location for the new jetport, Middletown, OH, was approximately 40 miles from downtown Cincinnati.
17. "Editorial: Down the Runway," *Dayton Daily News*, February 7, 1962; "Regional Airport Plan Debated; Cost, Distance Draw Criticism," *Cincinnati Post-Times*, February 23, 1962; "11-County Jet Airport Plan Debated," *Cincinnati Enquirer*, February 23, 1962.
18. "Patterson Backs Cincy 'Port Ties," *Dayton Daily News*, February 23, 1962; Dick Eben, "Regional Jetport Seen as Problem: Aid Priority Would Hit Local Plans – Starick," *Dayton Daily News*, April 5, 1962; "Editorial: Why Fear Area Jetport? It Would Benefit Us All," *Dayton Daily News*, April 6, 1962; "Jetport Fear Well-Founded, Chamber Official Declares," *Dayton Daily News*, April 6, 1962.
19. "Need Must Be Shown For Airport: FAA Chief Says It's Problem For Local Community," *Cincinnati Post-Times*, January 27, 1962.
20. "Butler County Names Two to Airport Group," *Cincinnati Post-Times-Star*, August 4, 1962; "Airlines Ask County to Delay on New Port," *Cincinnati Post-Times-Star*, August 10, 1962; "Airport Authority Created: Butler County O.K.'s Project," *Cincinnati Enquirer*, August 10, 1962; "Greene-Co Refuses Aid in Airport Plan," *Cincinnati Post-Times-Star*, August 15, 1962; "Airlines As Halt to Jetport Survey, Pending Own Study," *Cincinnati Enquirer*, August 15, 1962; "Montgomery-Co Votes Aid for Airport Study," *Cincinnati Post-Times-Star*, August 16, 1962; "Hamilton-Co Sure to Veto Airport Study," *Cincinnati Post-Times-Star*, September 5, 1962; "Editorial: What the Airlines Say," *Cincinnati Post-Times-Star*, September 5, 1962.
21. "Jetport Resolution Revised by County: Commissioners' Action Reaffirms $29,925 Contribution to Project," *Dayton Daily News*, October 10, 1962.
22. Fairbanks, "Responding to the Airplane," 173–179; Craig Lewis, "Ft. Worth Prepares Airport for Jet Era," *Aviation Week* 67 (October 28, 1957): 130.

23. Fairbanks, "Responding to the Airplane,"179–185.
24. Ibid., 184–187; "Texas Airport May Cover 21,000 Acres," *Aviation Week & Space Technology* 83 (October 4, 1965): 35; Robert Fairbanks, "A Clash of Priorities: The Federal Government and Dallas Airport Development" in Joseph F. Rishel, ed., *American Cities and Towns: Historical Perspectives* (Pittsburgh, PA: Duquesne University Press, 1992), 182.
25. Fairbanks, "Responding to the Airplane," 186–188.
26. Harold D. Watkins, "Miami Plans Large-Scale Training Airport," *Aviation Week & Space Technology* 89 (July 22, 1968): 39–42, 47, 49; Edward Hudson, "Miami Turning Swamp Into a Jetport," *New York Times*, November 30, 1968.
27. "Pollution of Everglades By Jet Planes Is Feared," *Sunday New York Times*, March 16, 1969; Nelson Bryant, "Wood, Field and Stream: Proposed Florida Jetport Considered a New Danger to Everglades Park," *New York Times*, April 4, 1969; "2d Hearing Set on Jetport in Florida Everglades," *Sunday New York Times*, June 8, 1968.
28. William M. Blair, "Transport Agency Is Accused Of Not Protecting Florida Park," *New York Times*, June 12, 1969.
29. Homer Bigart, "Naturalists Shudder as Official Hail Everglades Jetport," *New York Times*, August 11, 1969.
30. The "Everglades Jetport" remains open as the Dade-Collier Transition and Training Airport. Ken Kaye, "Airplanes and Alligators Mix at Remote Everglades Airport," *The Palm Beach Post*, February 28, 2010 http://www.palmbeachpost.com/news/news/state-regional/airplanes-and-alligators-mix-at-remote-everglades-/nL43J/ (Accessed 8 August 2014).
31. "Study Allegedly Opposes Jetport," *New York Times*, August 30, 1969; Felix Belair, Jr., "Jetport Is Termed A Threat To Park," *New York Times*, August 31, 1969; Christopher Lydon, "Gov. Kirk Joins Fight to Block Building of Everglades Jetport," *New York Times*, September 5, 1969; "Hickel, Kirk and Volpe Pledge to Aid Everglades," *New York Times*, September 11, 1969; Nelson Bryant, "Wood, Field and Stream: Fear of Water Pollution in Everglades Aroused Opposition to Nearby Jetport," *Sunday New York Times*, September 21, 1969; Robert Lindsey, "Irate Citizens Across the Nation Are Vigorously Resisting the Construction of Jetport," *New York Times*, December 26, 1969; Robert B. Semple, Jr., "Everglades Jetport Barred by a U.S.-Florida Accord," *New York Times*, January 16, 1970.

32. Bednarek, *America's Airports*, 151–177.
33. For examples of sites proposed as New York City's fourth airport and the debates surrounding those sites, see Glenn Garrison, "New Jersey Airport Plan Faces Hurdles," *Aviation Week* 71 (December 21, 1959): 30–31: _____., "Fourth N. Y. Airline Airport Disputed," *Aviation Week* 74 (May 1, 1961): 55, 57, 59; Joseph W. Carter "N. Y. Jetport Still Becalmed Amid Debate," *Aviation Week & Space Technology* 90 (May 19, 1969): 28–29; _____., "10 Carriers' New York Area Plan Discounts Need for New Jetport," *Aviation Week & Space Technology* 83 (November 1, 1966): 28. For more on the history of the effort to provide a fourth jetport for the New York region, see Michael N. Danielson and Jameson W. Doig, *New York: The Politics of Urban Regional Development* (Berkeley: University of California Press, 1982), 122–128.
34. Robert Lindsay, "State Takes Title to Land for Jetport," *New York Times*, August 14, 1971; Richard S. Kahn, "Steward Expansion Stirs Mixed Reaction," *Aviation Week& Space Technology* 95 (July 19, 1971): 28–29.
35. Quoted in Kahn, "Steward Expansion Stirs Mixed Reaction," 28–29; Richard Witkin, "Group Urges Limited Role for Stewart as a Jetport," *New York Times*, October 12, 1971.
36. Richard Witken, "$1-Billion Is Asked for Stewart Airport," *New York Times*, April 25, 1973;"Stewart Airport Seeks Jet Service, But Carrier Doubt Depth of Market," *Aviation Week & Space Technology* 129 (October 31, 1988): 92–94; Joseph Berger, "A Stealth Airport for New York; Lightly-Used Steward is New York's Stealth Jetport," *New York Times*, February 25, 1995.
37. Quoted in Robert R. Ropelewski, "Traffic Growth, Noise Wrack Los Angeles Airport Planning," *Aviation Week & Space Technology* 95 (November 15, 1971): 74; see also Robert Lindsay, "Irate Citizens Across the Nation Are Vigorously Resisting the Construction of Jetports," *New York Times*, December 26, 1969.
38. Ropelewski, "Traffic Growth," p. 74; "Palmdale Airport Recommended," *Aviation Week & Space Technology* 108 (May 8, 1978): 36; "History of Palmdale" http://www.lawa.org/pmd/pmd-History.cfm (accessed 3/19/2008).
39. Anne Randolph, "Completion of 1984 Slated for Los Angeles Upgrade," *Aviation Week & Space Technology* 118 (February 28, 1983): 33.

40. Steven P. Erie, *Globalizing L.A: Trade, Infrastructure and Region Development* (Stanford, CA: Stanford University Press, 2004), 172–185.
41. Ibid., 185–186.
42. Los Angeles World Airports (formerly the Los Angeles Department of Airports) owns and operates LAX, Ontario International Airport, the general aviation airport in Van Nuys, and the Palmdale airport site.
43. Ibid., 187, 225–226; "History of Palmdale."
44. Dan Weikel, "Palmdale City Council moves to develop shuttered airport," *Los Angeles Times*, April 3, 2009 http://www.latimes.com/wireless/avantgo/la-me-palmdale3-2009apr03,0,877428.story (accessed April 3, 2009).
45. Erie, *Globalizing L.A.*, 192–193.
46. Ibid., 194–197.
47. Quoted in Ibid., 197.
48. Ibid., 226.
49. Kenneth B. Ragsdale, *Austin, Cleared for Take-off: Aviators, Businessmen, and the Growth of an American City* (Austin: University of Texas Press, 2004), 169–173.
50. Ibid., 173–187.
51. Ibid., 188–196.
52. Ibid., 195–208.

Chapter Four: Airports for the "Jet Set": Expansion, Iconic Architecture, and Airport Malls

Jet airliners provided more space for passengers, but also required more space at the airport, especially in the form of longer runways. The new airplanes and the growth in passenger traffic also required new terminal facilities. And the airlines, spending millions to upgrade their fleets, expected local officials to invest in terminals that matched the glamor of the early jet age. Thus, cities wanting expanded jet service found themselves in a constant game of "catch up" as passenger traffic often grew faster than predicted, airlines rapidly shifted from piston-engine aircraft to jets, and both airlines and passengers came to expect a better airport experience.

The very earliest airport terminals were usually quite unimposing—often no more than a corner of an aircraft hangar. During the 1930s, more elaborate terminal buildings were constructed, but they remained fairly modest in size and with few exceptions utilitarian in function. In the post-war period, however, airlines began to sell not just airplane tickets, but an entire travel experience. Many airlines and their passengers complained that trips to exotic locations began and ended in small, often dirty, uninspiring airport terminals.[1] Further, as early as the 1950s, airport terminals, in many ways, began replacing railroad stations as a city's "front door." As a result, local officials became increasingly aware that a visitor's first impression might be made at the airport. In response, a number of cities, including St. Louis, Washington, and Los Angeles, began to invest in what they hoped would prove "iconic" airport facilities, and at

J.R. Bednarek, *Airports, Cities, and the Jet Age*,
DOI 10.1007/978-3-319-31195-1_5

New York's international airport, airlines competed with each other for most distinctive terminal.

The advent of airline deregulation in the 1970s created additional demands on airport design due to the growth of the hub-and-spoke system resulting in more multi-legged flights. As direct flights became less accessible, passengers found themselves spending more and more time in airports as they boarded their first flight at one airport and then changed flights at a hub airport before embarking to their final destination. To placate these stressed travelers, airport managers and airlines not only offered better restaurants and bars but also embellished the waiting areas and the portals leading to them with various kinds of shops, including ubiquitous chain stores, in which passengers could bide their time shopping. The US model for these new, more passenger/consumer friendly airports appeared in the early 1990s in Pittsburgh, based on European examples. It soon inspired imitators all over the country. By the early twenty-first century, airports across the USA had so transformed their facilities that they reminded passengers less of gateways to glamorous destinations and more of their home town shopping malls.

Terminals for the Jet Age: When Jet Bridges Were a Novelty

All cities faced the challenge of updating their airport facilities in order to be awarded jet service. Before deregulation, the Civil Aeronautics Board controlled the nation's air route structure. It not only determined where airlines could fly, but also ruled on what equipment it should and could use on those routes. In order to be certified for jets, airports had to meet certain minimal requirements, but as the jet age dawned, just exactly what those requirements would be were still under discussion. Also uncertain was to what extent federal aid—never as much as local airport officials wanted—would be available to upgrade for jet travel. While expanded runways and taxiways clearly fell under the categories of safety, efficiency, and convenience, new terminals did not. Basically, while federal funding became available for runways, taxiways, and noise-abatement, the terminal building—which for many passengers was "the airport"—remained the responsibility of local officials and the airlines serving their cities. Therefore, airside improvements (runways, taxiways, etc.) initially came before landside improvements (terminals, parking garages, etc.). Nonetheless, a number of cities rolled out the welcome mat for the new jet airliners with improved and expanded terminals. And after deregulation, as competition between cities for airline service increased, new terminal construction, some critics argued, took precedent over more runways.

In late 1958, Pan Am inaugurated its jet service from New York to London. Then, in early 1959, American Airlines flew the first domestic jet routes between New York City and Los Angeles. At that time, both the New York and Los Angeles airports had construction projects underway to more fully prepare for the demands of the new jet age. In anticipation of those demands, aviation experts had been debating the airport requirements for the dawning jet age. At a meeting of the American Society of Mechanical Engineers in late 1957, an official from American Airlines outlined many of the issues facing his company as it prepared to begin jet service in a little over a year. He argued that no airports served by American Airlines were ready to receive the new jets, either because of insufficient runways or inadequate terminal facilities, which meant that American jets would have to operate at less than maximum capacity. Other airline and aviation officials pointed to additional problems, such as the need for new fueling equipment and procedures, heavier towing equipment, and expanded cargo handling facilities with additional personnel. And they worried about how efficiently to board and deboard the larger number of passengers as the new jets carried nearly twice as many passengers as the largest piston-engine airplanes.[2]

Federal government officials also worried about other problems that might emerge in the jet age. Just as airlines began planning for the new era, Congress was in the process of creating a new, independent Federal Aviation Agency. Among the responsibilities of the new agency would be the modernization of the nation's airways and airports to meet the demands of new aircraft technology. Even before passage of the legislation, the soon-to-be-replaced CAA created an interim Airways Modernization Board. It embarked on a program of research and development aimed at identifying what was needed for a fully modern, jet-age airway and airport system, though some progress had been made in providing cities with information they needed to prepare immediate airport improvement plans. For example, following the certification process for the first generation of US jet airliners, Congress approved $200 million for improved runways and taxiways at major airports, and federal agencies distributed information about jet runway length and strength requirements.[3]

Some local airport officials, wishing their cities to be among the first to boast of jet service, immediately embarked on airport improvement programs, including the construction of new terminal facilities. For example, the PNYNJ had already begun improvements at Idlewild, and Los Angeles was just launching its expansion plan. In addition, programs were also underway at Dallas, Miami, and Cleveland, among other cities, with Chicago and Minneapolis-St. Paul anticipating the start of their own improvement programs in the near future. However, it was clear that jet service would begin before all the new ground facilities, especially the new jet terminals, would be ready. An aviation official warned that the new jets would be operating out of "basic facilities only. The frills that will give us complete efficiency will come later – much later."[4]

Many agreed with that assessment. For example, in March 1958, an article in *Aviation Week* stated that only one airport would have a terminal (San Francisco) fully ready to receive the new jets. Though American Airlines was scheduled to begin its transcontinental jet service between New York and Los Angeles in January 1959, its terminal in New York would not be completed until June 1959. Los Angeles's new airport was not scheduled to be finished until the early 1960s. At all the other cities where American anticipated bringing jet service—including Chicago, Boston, Baltimore, Detroit, Fort Worth, Dallas, St. Louis, Phoenix, and Tucson—the airline would have to operate out of interim facilities.[5]

The aviation press was not the only forum for discussion of new jet age airport requirements, though. The popular press also weighed in, usually with a more optimistic tone. For example, just as American Airlines was about to begin its jet service, an article in *Time Magazine* profiling the airline's president, C.R. Smith, highlighted projects underway at 40 US airports to build the necessary ground facilities, including extended runways. The article also noted the airline's expenditures on terminal upgrades, including $14 million spent on its new facility at Idlewild, as well as new larger hangars. Passengers, the article forecast, would "wait for their flights in comfortable, soundproof lounges, [and] board the jet on a single level through telescopic covered passageways that shoot out to the plane's two doors." President Smith also asserted that the jet age would involve faster ticketing and baggage handling as well as better food and lower fares. The article admitted that many of the needed changes had not yet fully materialized, but that such changes were "inevitable, simply because the jet age demands them."[6]

In reality, three months after the *Time* article appeared and just as American began domestic jet service, the FAA reported that only about 25 airports in the USA were capable of handling the new jet aircraft carrying a "reasonable load." None could handle them at "maximum load." The FAA estimated it would cost close to $1.3 billion to meet the needs of those airlines shifting to jet service as runways needed to be extended and secondary runways constructed. The new agency also determined existing terminals were not only inadequate for the jet age, but they were barely meeting the current needs. And, as *Aviation Week* reported, the inauguration of jet travel would result in a rapid increase in the number of people traveling by air. In 1958, according to the article, the commercial airlines carried 49 million passengers. By 1960, it was projected that the number would increase to 66 million. And by 1970, the number could be as high as 118 million (a conservative estimate, it would turn out).[7] The FAA report concluded that airport development lagged due to a lack of governmental funding at all levels, federal to local. And a spokesman for airport operators opined that the federal government should allow cities to use aid funds for terminal construction and improvements, but runways and taxiways would come first.[8]

Time Magazine, however, remained upbeat about steps taken toward establishing adequate jet age airports. In a 1960 article on "airport cities," it highlighted the start of major airport projects not only in the USA but around the world. It detailed the ways in which the new terminal facilities would be larger and more comfortable than existing terminals. While it described many of the airport improvements as "functional," it noted that some of the projects included "dramatic" new designs for the jet age. In addition to Idlewild and Dulles, the article singled out the new terminal under construction at O'Hare Airport in Chicago, soon to replace Midway as the city's major airport. The airport itself would be one of the largest in the world and would have three terminal buildings to serve passengers.

The article also applauded promising plans for moving passengers and baggage quickly and smoothly through the new terminals. United Airlines was experimenting with what would become one of the more popular innovations of the jet age—the so-called jet bridge—that eliminated the need for passengers to walk in all kinds of weather across the noisy and smelly tarmac and onto a boarding ladder at the plane. Instead, the jet bridges consisted of covered walkways that telescoped out from the terminal to the plane's passenger doors. And at Chicago O'Hare, a conveyor belt would carry baggage into a more passenger-friendly baggage claim area.[9]

The *Time* article concluded with a somewhat rosy statement from the head of the FAA, Elwood Quesada, about airport design. He said that the federal government had designed the new airport it had built just outside Washington, D.C., Dulles International Airport, to meet not just the needs of the 1960s, but also the 1970s as well. Airport designers, he opined, should think well into the future. "Not looking far enough ahead," he said, "is one of the errors we've been making through the history of commercial aviation." But now, we have "forecast the requirements and are not indulging in building for today. We are building for ten years, twenty years, fifty years from now."[10]

But predicting the future requirements of air travel proved far more difficult in practice than in theory. City after city built new terminals and extended runways with the idea that these would meet demand well into the future. For example, Atlanta's municipal airport underwent a major expansion in the early 1960s. It included new and expanded runways as well as a new terminal building, replacing a World War II-era converted hangar that had served as the city's initial post-war terminal. However, the airport terminal was overwhelmed as soon as it opened in 1961. It was designed to handle 6 million passengers per year; in the first year of operation, it handled over 9 million. Within three years, the city had to begin planning another major expansion of the airport, including additional terminal facilities.[11]

The doubling and redoubling of the number of passengers through the 1960s and 1970s overwhelmed many relatively new facilities across the country. Los Angeles, San Francisco, Washington, and New York had all built new or expanded airports by the early 1960s, but at mid-decade, officials at each of those airports were scrambling to find ways to expand capacity.[12] Chicago's O'Hare Airport, completed in 1963, had exceeded the number of passengers expected by 1970 as early as 1965. And few airports had adequate space for the automobiles those passengers used to travel to the airport. To increase the handling capacity of the existing facilities, airports and airlines experimented with new ways to handle ticketing and baggage check-in. Further, to accommodate the new, larger jets, airport terminals added long "finger concourses" stretching out from the main facility. While these provided more room for the jets, they also meant longer walks for the passengers. For example, at Atlanta, Miami, and Chicago, passengers might have to walk as much as a half-mile from the ticketing area to the gate.[13] And while a number of cities explored the idea of dealing with the capacity issue by building completely new airports, the actual number of new facilities was quite small. Between the early 1960s (after Dulles opened) and the early 1970s, only Houston and Dallas-Fort Worth in Texas and Kansas City, Missouri, dedicated new airports. High costs, an economic downturn in the early 1970s, and local opposition due to environmental concerns, especially the persistent noise problem, all worked to effectively shelve most plans for completely new airports and delayed many plans for airport expansion.[14]

The capacity issue continued to challenge airport officials and airlines through the 1970s into the 1980s. But during the 1980s, deregulation set off a sharp rise in the number of passengers from just over 200 million in the mid-1970s to 410 million in 1987, a boom that led to extensive terminal construction.[15] Deregulation freed airlines to set their own route structures. That, in turn, set off a level of competition between cities for airline service probably not seen since the early days of commercial aviation. The evolving hub-and-spoke system played a part in the boom by promising the reward of increased passenger traffic to airports selected as hubs and also encouraged airport officials to rethink their lease arrangements with airlines. Many moved away from the model of long-term leases to shorter-term agreements in order to secure more flexibility in attracting new airlines to their markets.[16]

Under these circumstances during the 1980s, airport improvement projects started or reached completion in cities across the country. As noted, Los Angeles, as part of its pitch to host the summer Olympics in 1984, included a new international terminal as well as improvements in its domestic terminals. At the same time, Chicago secured approval for a major expansion at O'Hare, Boston started to modernize and expand Logan's Terminal C, the Dallas city council approved a three-year $28 million improvement at Love Field, and Orlando dedicated a new international concourse that doubled the airport's capacity. Smaller cities, such as Milwaukee, Omaha, and Cedar Falls (Iowa), also participated in the airport improvement sweepstakes during the deregulation bonanza of the 1980s.[17]

Iconic Architecture: New City Symbols for a New Air Age

The updating of terminal facilities in the 1960s and afterward primarily aimed at serving the domestic market. Increasingly, though, as more Americans began to travel abroad and more foreigners began to travel to the USA by air, US airports found themselves in competition with international airports around the world. If they were not strictly in competition for passengers, they were in a competition for the biggest, best, and most spectacular terminal facilities. Cities interested in attracting more international airlines and travelers especially worked to improve their terminals as their "front door" to the world.

During the 1920s and early 1930s, the USA lagged considerably behind Europe in terms of the quality of airport architecture. Though the USA caught up with Europe in terms of airplane technology—a lead it had lost just before World War I—Europeans took the lead in developing the first modern airports and the distinctive buildings that came with them. As in the USA, airports in Europe began as simple facilities—large grassy areas with a few utilitarian buildings. By the 1920s, however, state-sponsored airlines pioneered international air travel in Europe. The airports and terminal buildings constructed to serve these airlines were designed as symbols of national pride. European architects, in many ways, invented the modern air terminal, and the structures at airports in London, Paris, Berlin, and other European capitals became the standard.[18]

The USA and Europe, however, seemed equal in terms of the fantastic schemes dreamed up for airports and aviation in the 1920s and 1930s. Architects, artists, engineers, and other would-be visionaries in both places imagined airports of the future. Many, including Frank Lloyd Wright and Le Corbusier, sought to integrate aviation and their visions of the ideal city. Skyscraper architects, including Lloyd Wright, son of Frank Lloyd Wright, sketched designs which combined tall buildings, airplanes, and dirigibles. Though futuristic ideas never completely disappeared, among more serious architects and designers a greater sense of realism took hold by the late 1930s.[19]

Reflecting this more practical sensibility, architects in the USA designed airport projects just before World War II in New York City and Washington, D.C., that in many ways set new American standards for airport terminals. Construction began on Municipal Airport No. 2 (later LaGuardia Airport) in 1937. The firm of Delano and Aldrich designed the new airport, which provided separate terminals for seaplanes and land planes. The Marine Air Terminal, still in use today after being redesigned to service land planes, featured a large, domed interior with a central skylight and an exterior decorated with a terracotta frieze of flying fish. The separate landplane terminal included a restaurant and an observation deck and many other design features now standard, including the vertical separation of the ticketing area and the departure gates—passengers arrived at the airport on an upper level and then proceeded to the lower level for boarding the aircraft (Fig. 1).[20]

Shortly after the completion of the New York airport, construction began on a new facility to serve the nation's capital. The new "National

Fig. 1 Marine Air Terminal, LaGuardia Airport with exterior frieze of flying fish. Library of Congress, Prints & Photographs Division, HAER, HAER NY, 41-JAHT, 1–1

Airport" was located on landfill across the Potomac River from the District of Columbia. The design of the terminal building reflected a unique combination of Art Deco and the colonial style of George Washington's home at nearby Mount Vernon. Instead of organizing the terminal vertically, as had been done at LaGuardia, the Washington terminal separated passengers horizontally—arriving passengers entered the south end of the building while departing passengers exited the north end. The new terminal building opened to nearly unanimous acclaim.[21] In design and scale, the new terminals in New York and Washington foreshadowed the eventual use of air terminals as symbols of city identity and stature in the USA.

Just before the dawn of the jet age, St. Louis, Missouri, was one of the first US cities to receive acclaim for its post-war terminal. When the city of St. Louis's airport commission decided to build a new terminal building in 1953, it hired the firm of Hellmuth, Yamasaki, and Leinweber. Design of the terminal fell to one of the young partners in the firm, Minoru Yamasaki. At that time an unknown, Yamasaki would soon make his mark not just on St. Louis, but on American architecture generally. Following his work at the airport, the housing authority in St. Louis hired his firm to design the now infamous Pruitt-Igoe public housing complex. He also served as the lead architect on the twin towers of the World Trade Center in New York City.[22]

Yamasaki's St. Louis airport terminal, though conceived and constructed before the commercial jet age, in many set the standard for a number of later iconic terminal structures. The building, designed to be expanded, consisted originally of "three pairs of intersecting barrel vaults" reaching a maximum height of 32 feet and creating an interior space of 412 feet long. The barrel vaults terminated in eight floor-to-ceiling walls filled with windows. These allowed light to flood the building during the day, while interior lights shining through the windows made the building exceptionally visible at night.[23] The new terminal opened in 1956 (Fig. 2).

Fig. 2 Lambert St. Louis Terminal, circa 1950s (Courtesy Lambert St. Louis International Airport)

The St. Louis airport design, with its use of concrete vaults, clearly influenced the work of Eero Saarinen. Saarinen was born in Finland in 1910, the son of a celebrated architect. His father, Eliel Saarinen, moved the family to the USA in 1923. The young Saarinen attended Cranbrook Academy and in 1934 graduated from the Yale School of Architecture. By the late 1930s, he launched a successful career as an architect, and in the late 1950s, Saarinen received commissions to design what became two landmark US terminal buildings—the TWA terminal at Idlewild (later JFK) airport in New York City and the terminal at Dulles International Airport.[24]

Saarinen's company received the commission for the TWA terminal at Idlewild in 1956, the first airport building the firm designed. Rather than the single terminal building for the airport, though, it would stand as one of a number of terminal buildings on the site because the port authority and its architectural firm, Harrison and Abramovitz, had devised a unique airport master plan for the city's new international airport that called for not one but for a series of terminal buildings—each scheduled for design and construction by the airline using it.[25] The terminal city idea amounted to a radical departure from the familiar practice of creating a single building to serve all airlines. This break from the customary design came at a time of transition in commercial aviation. The jet age loomed on the horizon, quite visible to airport planners and engineers, while at the same time, airlines and airports already struggled to keep up with mushrooming levels of air travel. Eventually, New York's terminal city consisted of nine terminal buildings.[26] Therefore, Saarinen's first airport commission was part of a somewhat radical and controversial project.

Further, the design of airport terminals was an evolving art in the 1950s. Because the terminal would serve only one airline, it potentially allowed for more rapid processing of both arriving and departing passengers. To fully take advantage, Saarinen and his firm began a careful study of the movement of people, baggage, and airplanes in and around the terminal. Saarinen sought a design that would make the movement of each most efficient. His proposed solutions to some of the problems eventually became standard practices. For example, as he envisioned it, passengers would leave their baggage at the main check-in facility. From there the baggage would move to the planes on one level while passengers proceeded to the boarding area via a covered bridge with moving sidewalks. Arriving passengers would also use moving sidewalks to travel from the gate to the area containing the new baggage carousels. His vision, however, was not fully realized at the time. The baggage carousels were not fully automated, for example, and the original passenger bridges did not include the moving sidewalks.[27]

Shortly after receiving the commission for the TWA terminal, Saarinen and his company was hired to design the terminal building for the new Washington area airport. Congress first authorized construction of a new airport to serve the national capital in 1950. However, controversy over site selection repeatedly delayed the project. Congress sought action again in 1956 in the face of growing air traffic congestion at Washington National Airport. In 1957, Eisenhower assigned Elwood P. Quesada, then a special assistant to the president, to study the proposed sites and make a recommendation. After evaluation and public hearings, Quesada recommended a site near Chantilly, Virginia, 26 miles west of Washington, D.C. Eisenhower announced the decision in January 1958, and work on the design and construction of the new airport began in May of that year.[28]

Saarinen began designing the new structure in 1958, the year before the introduction of commercial jet air travel in the USA. Therefore, the new Washington Airport, named in honor of Secretary of State John Foster Dulles, would be among the first airports and terminals designed and built for the jet age, and it was the first completely new airport designed with jets in mind.[29] It would also function as the international air gateway for the nation's capital. Saarinen wanted it both to reflect the position of the USA in the world and demonstrate a connection to the federal architecture so prevalent in the nation's capital. The building would also stand as the entrance to what was then the nation's largest airport in terms of area. At 15 square miles, it was twice the size of the new international airport in New York City (Fig. 3).[30]

In response to the design challenges, Saarinen created a single massive terminal building. He positioned the building carefully so that it could be viewed from the access road, from the elliptical road leading to the terminal, and from the parking lot in front of the building. The terminal was originally 600 feet long and 150 feet wide. The concourse ticketing area was a single large room. Departing and arriving passengers were handled on two different levels (Fig. 4).[31]

As noted, one of the major challenges increasingly facing all airport designers was the distance passengers had to travel within the airport to either board the aircraft or get from the aircraft to the terminal lobby and baggage claim. Also, the larger size of the newer airports in particular meant longer taxiing routes for the airplanes, which cost time and fuel. The massive size of the Dulles Airport and the new jets that would serve

Fig. 3 Exterior Dulles International Airport. Library of Congress, Prints & Photographs Division, Balthazar Korab Archive at the Library of Congress, LC-DIG-krb-00760

it promised to further complicate the situation. Saarinen had proposed moving sidewalks in his TWA design. At Dulles, to move passengers to and from the terminal building and the awaiting airplanes, Saarinen devised another unique solution—the mobile lounge. Unlike the moving sidewalks, which eventually would become features in many large airports, the mobile lounge was not adopted much outside of Dulles. Saarinen based his mobile lounge idea on the European practice of moving passengers to planes using buses. Saarinen, however, envisioned a far more complex vehicle to do the job. The mobile lounges were essentially

Fig. 4 Interior Dulles International Airport. Library of Congress, Prints & Photographs Division, Balthazar Korab Archive at the Library of Congress, LC-DIG-krb-00788

waiting rooms on wheels. Departing passengers would enter the mobile lounges after a short walk from the main concourse. The vehicles, which not only moved horizontally but could adjust vertically, carried the passengers directly to their waiting airplanes. Arriving passengers went directly from the airplane to the mobile lounge, which then docked at the terminal building. The airplanes, thus, avoided the long taxi from the runways to the terminal (Fig. 5).[32]

Though the mobile lounge idea did not catch on elsewhere, the terminal building became something of the prototype for the modern airport terminal. Architectural critics, though, disagreed as to which terminal building—TWA or Dulles—represented the best of Saarinen's work. While Saarinen considered the Dulles building his best work, the TWA building captured the affection of the critics.[33] Revolutionary at the time of its construction, the TWA terminal proved inflexible in meeting the changing demands over the last half of the twentieth century, however. Its future was further complicated following the merger between TWA and American Airlines in 2001. With the demise of TWA, the PNYNJ decommissioned the building. New York preservationists, however, rallied

Fig. 5 Mobile Lounge, Dulles International Airport. Library of Congress, Prints & Photographs Division, Balthazar Korab Archive at the Library of Congress, LC-DIG-krb-00771

to prevent its demolition. Rescue came with expansion plans announced by discount airline JetBlue. JetBlue began operations out of JFK in 2000 and by 2005 was the airport's busiest carrier, operating out of Terminal 6. In late 2005, however, it announced plans to build a new 26-gate terminal behind and surrounding the old TWA building, by then known as Terminal 5. The new JetBlue terminal opened in 2009. How the old TWA building will be used was still evolving with plans in late 2015 having it open as a luxury hotel in 2018 (Figs. 6 and 7).[34]

The west coast also witnessed the construction of a now iconic airport building. The city of Los Angeles, California, experienced rapid growth

Fig. 6 Exterior TWA Terminal at JFK. Library of Congress, Prints & Photographs Division, Balthazar Korab Archive at the Library of Congress, LC-DIG-krb-00599

in air traffic in the post-war period. To meet growing demand, the city announced plans to vastly expand its airport as early as 1953. The initial plans called for the creation of a new, expanded system of runways as well as a centrally located, round, domed terminal building. The plans were said to have a "Buck Rogers look." The city's Airport Commission placed a $33 million bond issue before the voters.[35] However, that and subsequent bond issues failed. The city also considered issuing revenue bonds to finance the project, but found that investing companies and lending institutions were not interested in underwriting them. In the meantime, both passenger and cargo traffic at the airport continued to grow dramatically, straining the capacity at the airport's terminal.[36] Finally, in 1956, the airport commission gained voter approval of a bond issue and in November 1957, the same month that Boeing rolled out the first, flight-test version of its new commercial jet aircraft, Los Angeles unveiled its new plan for airport expansion.[37]

Fig. 7 Interior TWA Terminal at JFK. Library of Congress, Prints & Photographs Division, Balthazar Korab Archive at the Library of Congress, LC-DIG-krb-00611

Construction began in late 1957 and was scheduled in three phases, with the final phase scheduled for completion in January 1960 (actually completed in 1962). The "Buck Rogers" domed terminal building from the 1953 plan was gone. The 1957 plan, however, included an even more futuristic structure, one which would become the unmistakable symbol of the new LAX. Whereas TWA in New York City and the federal government at Dulles built iconic terminal buildings, Los Angeles built instead a distinctive building housing a restaurant and observation deck. Architect William Pereira designed the so-called "Theme Building," completed in

1961. Unlike the TWA and Dulles terminals, the Los Angeles structure evoked not atmospheric flight, but space flight. Though designed the year Sputnik inspired both fear and awe in the minds of the American public and completed the year Alan Shepard became the first American in space, the 1953 movie "War of the Worlds," not the emerging space race, inspired the building's architecture. The architect's brother served as the film's art director, and the Theme Building was said to have been designed to evoke the image of the Martian invaders from that film.[38]

The Theme Building came to stand for the hopes and aspirations many invested in what they saw as the modern and future-oriented city of Los Angeles. Like the TWA terminal, it faced a number of challenges in the first decade of the twenty-first century. The attacks of September 11, 2001, resulted in the closure of the building's observation deck. After a large piece of the building's stucco exterior broke off and hit the roof of the restaurant, the popular Disney-designed drink spot closed for eight months and an extensive renovation began. The $12.3 million project was completed in 2010, resulting in a refreshed and earthquake-proofed structure.[39] With the addition of new security measures, the airport reopened the building's observation deck in July 2010 (see cover illustration).[40]

By the 1970s, whatever momentum there had been toward building distinctive airport buildings had all but halted. Architectural critic Alastair Gordon blamed the trend toward more functional and prosaic designs on a number of factors. First, there was the challenge of simply trying to keep up with a passenger load that was doubling every decade from 1958 to 1987. Second, security demands brought by the wave of hijackings that began in the late 1960s resulted in terminals that resembled bunkers, with thick concrete walls and perimeter barriers. The free flow of passengers gave way to an architecture aimed at funneling people through a single, secure entry point.[41]

In addition, the poor economy during much of the 1970s also made expansive and expensive airport plans far more difficult to finance. Neither cities nor the airlines were anxious to bankroll such projects, except when absolutely necessary. Where projects were undertaken, the plans tended toward functionalism and cost efficiency as local officials generally decided to expand existing facilities as much as possible rather than building new. Airlines and airport operators also experimented with ways to make existing facilities more efficient, but even such large cities as Chicago, Boston, and Philadelphia all saw airport improvement and expansion plans scaled back or re-worked as officials and the airlines could not come to agreement over financing.[42] The economic woes continued into the early 1980s. High interest rates and a Reagan Administration plan to "defederalize" the nation's largest airports all made bond issues difficult to sell.[43]

But the decades after the 1980s witnessed a world-wide wave of airport construction. Particularly in Asia and the Middle East, a number of nations supported the construction of grand, gateway airports. Massive airport projects included Kansai International in Osaka, Japan; Sepang International Airport in Kuala Lumpur, Malaysia; the New Seoul International Airport in South Korea; and Hong Kong's Cep Lap Kok. Big projects, though, were not limited to Asia and the Middle East. In Great Britain, Stansted opened as London's third airport in 1991 with a new terminal designed by celebrated architect Sir Norman Foster. In the USA, United Airlines financed a new terminal at Chicago's O'Hare, and the PNYNJ developed a plan for a massive overhaul of JFK International.[44] In the USA, though, no new major commercial airport had opened since the Dallas-Fort Worth airport in 1974. That changed in 1995 with the dedication of a new airport serving the city of Denver. It opened as the largest airport in the country and with a terminal designed to serve as a symbol for the city.

At least initially, Denver International Airport was known mostly for its cost, its massive land area (50 square miles), and the never-functioning automated baggage handling system. However, it also has been recognized for its unique terminal. From the beginning, airport proponents envisioned the facility functioning as a new symbol for the city of Denver. They wanted it to be as dramatic and memorable as the Eiffel Tower in Paris or the Arch in St. Louis. The most striking feature of the terminal—its tent-like roof structure—was actually a late addition. The original design called for a massive glass and steel roof, one that was reminiscent of the great railroad terminals of the past. City leaders, however, did not believe that the original design would make the powerful statement they envisioned. Therefore, the city turned to a local architectural firm, Fentress & Bradburn, to design an alternative roof structure that would make a distinctive architectural statement. Given three weeks to develop a cost-efficient and buildable design, the firm proposed the tent-like, Teflon-coated fabric roof.[45]

Denver's was not the first airport terminal to use such materials. The main terminal at Saudi Arabia's King Abdul Aziz Airport had a roof structure designed to look like a tent, using the same Teflon-coated fabric. The multi-peaked roof at Denver, however, was not designed to evoke a tent or tents, but the nearby Rocky Mountains, with each peak soaring 120 feet above ground. The fabric allowed for light to filter into the terminal during the day. Glass panels in the peaks as well as around the exterior of the roof also allowed in direct sunlight and, at night, interior lighting in the building combined with the translucent fabric made the roof seem to glow. The new roof was expected to actually save costs over the original design, and the city hoped to use the expected savings to upgrade some flooring materials and double the size of the parking structure. As with much else at the Denver airport, however, the costs rose tremendously. Nonetheless, in the end, the Denver terminal cost less per square foot ($100) than United's new terminal facility in Chicago ($370).[46]

But changing the roof, while maintaining the original design of the interior, compromised the essential goal of having a completely unified design throughout the terminal. And while the terminal's roof might evoke the Rocky Mountains, the terminal provided neither a view of the actual mountains nor of the Denver skyline. On the other hand, the city did not compromise, at least in terms of spending, on the airport's public art program. With a budget of $7.5 million, the Denver airport ranked at the time of its construction as the largest such program in the country. Art critics both praised and panned the art collection, but, and perhaps more importantly, passengers who used the airport gave the art high ratings.[47]

Competition between cities in the USA for air traffic resulted in the design and construction of a number of unique terminal facilities. By the early twenty-first century, though, international competition also figured into the equation as architecturally distinct terminals graced airports from Hong Kong to Dubai, Japan to Bangkok. When *Travel and Leisure* named the world's most beautiful airports in January 2010, only two US airports made the list—the TWA terminal building at JFK (designed in the 1950s) and Denver International Airport. Five of the most beautiful were in Asia, three in Europe, two in South America, and one in Africa.[48] Though unique terminals seemed less a national priority in the USA, nonetheless a number of cities, and even the federal government, at one time or another invested in terminals designed to serve as both transportation facility and as symbol.

The Malling of Airports

Europeans pioneered the modern airport terminal, and it was also from Europe that the USA got the idea of turning airport terminals into shopping malls. The mall-like design and operation of airport concessions came to the USA through Pittsburgh and reflected the international movement toward the privatization of airport operations to make them appealing to passengers. Making airports more appealing became especially important in the USA after deregulation as the new hub-and-spoke system created long wait times for air travelers with connecting flights at the new hub airports. The BAA experimented with its new mall concept at Heathrow in order to bolster airport revenues. Duty-free shopping had long comprised an important source of revenue for airlines at Heathrow, as elsewhere, but a new European Union-mandated ban on tax-free shopping prompted the BAA to expand and diversify retail opportunities at Heathrow on the landside of the airport before passengers reached customs. And it worked. Very quickly fees from retail concessions provided 65 % of the airport's revenues.[49]

In the USA, airport terminals had long housed a variety of shops and restaurants. In fact, in the early years, airport operators looked for any opportunity to generate revenue. Swimming pools, tennis courts, oil wells, observation decks, and pay toilets all contributed to airport revenues. When air travel was dominated by the more affluent, airports often boasted of high-end restaurants and elegant cocktail lounges or nightclubs.[50] But the shops and eateries were fairly limited in numbers within the airports. One could get a haircut, a shoe shine, buy a newspaper, or wait out a flight over a cocktail, but airports did not yet resemble the expansive shopping malls beginning to make their marks on America's suburban landscape in the 1950s and 1960s. That changed in the 1990s.

Pittsburgh, Pennsylvania, the city and its metropolitan region, was hard hit by deindustrialization. Local leaders adopted many strategies in the 1980s to spur new economic development, including a plan to rebuild the city's airport. The Allegheny County Airport Authority, which owned the airport, had long bought surrounding land to deal with the noise issue and to provide for expansion (see chapter "The Broad Problem of Airport Noise: Airports, the Courts, the Federal Government, and the Environment"). By the time the county embarked on the improvement project, Greater Pittsburgh International Airport was second in total land area only to the Dallas-Fort Worth airport, having grown from 5000 acres to 12,000 acres. Further, during the 1980s, Allegheny Air, the main carrier at the airport, grew through merger and acquisition into USAir and made Pittsburgh one of its hub airports. The heavy hub-and-spoke traffic strained the capacity of the old, 54-gate airport facility and provided the tipping point for the campaign to build a new terminal.[51]

The new terminal was designed to have three distinct parts. A "landside" section was designed to hold the ticket counters, baggage claim, and rental car counters. A second section held 25 commuter flight gates as well as the airport's security checkpoint. The third section was the new X-shaped main concourse, constructed with 75 gates with room to expand to 100. Moving sidewalks and passenger trams in the building expedited transit and connected the three parts, while the mid-field location of the terminal, along with the X-shape of the main concourse, allowed for shorter taxi times for arriving and departing flights.

County and airport officials emphasized the efficiency of the airport's arrangements for handling traffic and hoped that it would spark economic development in the relatively undeveloped territory around the airport. Like the city's business boosters, they predicted that the new airport's super-efficient handling of traffic and the new 7.5 mile expressway link to the city would attract high-tech industries, then the nation's most rapidly growing economic sector, to the area and give it an insurmountable lead in that field over its urban eastern US competitors. For these reasons, the county and state governments invested millions of dollars of the total cost of the project, which rose to $1 billion.[52]

The new airport opened in September 1992, and even before its debut, many observers began referring to it as "revolutionary." Many users of this adjective referenced either the general design of the entire ensemble of parts or the price tag, or both. But once the terminal opened, much of the sense of the airport's revolutionary character focused on the mall-like character of the food courts and stores located in the center of the X-shaped terminal, and the term did in many ways accurately portray the structure and management of the concessions at the airport.

The BAA brought the practice it had pioneered at Heathrow directly to the USA after the Allegheny Port Authority decided to contract the management of concessions at the Pittsburgh terminal to a private company. The BAA outbid four American companies and won a 15-year contract. Previously, stores operating in US airport terminals had been granted essentially monopoly contracts and could charge high prices because they lacked competition. The Pittsburgh model changed all that. The BAA granted no monopoly concessions. Instead, it contracted with a number of franchises to provide food or other goods—three separate companies, for example, held newspaper concessions. Further, the concession contracts emphasized what BAA called "street pricing"—concessionaires were not to charge highly inflated prices for their goods compared to what they charged in their non-airport locations. In addition, the airport design put all the airport concessions just beyond the single security checkpoint. All departing passengers, therefore, passed through the shopping area on their way to the gates and connecting passengers waiting for their flights did not have to pass through security in order to access the shops. And in the days before September 11, 2001, individuals coming to the airports with departing passengers or to pick up arriving passengers could also pass the time in the shopping area. Finally, the BAA not only managed to fully rent all the retail space before the terminal opened, but more than half of the companies opening shops at the Pittsburgh airport had never operated at an airport before. This provided Pittsburgh International a unique and wider mix of retail options than any other airport in the USA.[53]

But not for long—after Pittsburgh, the BAA installed its mall concept in Indianapolis when it took over management of that city's airport system. A *New York Times* article announcing the contract emphasized that what travelers would notice most about the change in management would be "more name-brand stores and concessions at the airport."[54] Soon, whether managed by BAA or not, airports all over the country responded to the Pittsburgh model. In 1997, for example, the PNYNJ completed a refurbishment of the main terminal at LaGuardia Airport. Though other improvements were also part of the project, most attention focused on the new mall, filled with shops and food vendors, on the second floor of the terminal in a new 45 foot wide, three-story atrium where once stood a statue of Fiorello LaGuardia.[55]

The announcement of the opening of the LaGuardia airport mall claimed that it represented a new way of thinking about airports. The thinking was not as new as the port authority claimed, but it was clear that airports were no longer viewed simply as a place to handle air passengers. The remodeling focused on bringing name brand stores to the airport in an effort to increase the popularity of LaGuardia among passengers accustomed to international standards of airport operation. As New York City Mayor Rudoph Guiliani insisted, world class airports in the late 1990s should offer world class shopping opportunities for air travelers. There were those who worried about the viability of the new mall at La Guardia, however, as it was not a hub airport, so it did not have connecting passengers with shopping time between flights.[56]

Despite challenges, by the late 1990s, both hub and non-hub airports were turning to retail as a way to entice travelers to use their facility. And retailers were seeing the advantages of airport locations. Inventory, staffing, and long hours of operation all challenged retailers to rethink their traditional shopping mall strategies. However, when successful, many retailers found that they could make far higher profits per square foot of retail space at the airport than at a traditional mall. And both airport officials and retailers were learning lessons. For example, shops located past the security check point tended to do better than shops located before the check points. Also, as suggested above, hub airports tended to be more profitable than non-hub airports. Therefore, before September 11, 2001, the Pittsburgh model—with all shops past the single security check point—made the most sense. On the other hand, the new terminal at Washington Reagan National Airport seemed to provide the example of a less than ideal retail environment. Not only was the airport not a hub, but the major shops were located before the security check point. Few passengers, once past security, wanted to exit the secure area in order to shop. Companies with stores in both airports reported that their Pittsburgh stores were doing far better than their Washington stores.[57]

The expansion of retail and other services at airports did not come without controversy. By summer 2001, both air travel and delayed flights reached record levels. Across the country, critics noted, billions were being spent on airport expansion programs, but the expansion took place mostly on the landside rather than the airside. Only six runways were added at major airports in the 1990s. In the early 2000s, airports drew up plans for 15 additional runways, but aviation experts charged that runway expansion itself lagged far behind the increase in traffic. At the same time, and despite the boom in the 1980s and 1990s in airport retailing, many argued for the addition of even more retail outlets at US airports. Even though profits from retail grew steadily at US airports, foreign airports did even better. European airports, for example, generated 20 to 25 % of their revenues from retail sales, while the US figure came in at 10 to 12 %.[58]

Then, as noted, the September 11, 2001, attacks undercut the retail environment at US airports. Initially, retail sales fell twice as much as passenger loads (40 % vs. 20 %), and the retail slump persisted as airport officials restricted access to areas beyond the security check points to ticketed passengers. But even airports with shops before security suffered from significant drops in sales. With passengers focused on making it through extremely long lines at security, they would neither pause to shop before getting in line nor would they exit the secure area again in order to shop. One airport designer even opined that the new airport malls might belong to another era in the wake of the terrorist attacks.[59]

Yet by early 2003, airport retail sales seemed on the rebound. The security protocols put in place after September 11, 2001, encouraged passengers to arrive at the airport hours before their flight time. Long security lines could take up much of the added time, but when they did not, passengers were left with increased waiting time before their flights. And to pass the time, passengers shopped. At Pittsburgh International Airport, passenger volume declined 14 % in the last quarter of 2001, but retails sales fell by only 11 %. As US air travelers grew accustomed to allowing for more time at the airport, they also began to spend more money while there. Food sales, especially ice cream, climbed as airlines increasingly cut back on the food served on the airplane. Book sales also rebounded as passengers spent the time waiting for their flights reading. Retailers, who catered to business travelers increasingly dependent on their laptops and other electronic gear, opened shops offering charging stations, fax machines, and modems. Airport hotel chains also provided new amenities, including low-cost workout rooms or fitness centers. And as more families began to travel together more airports added play areas to keep youngsters happy as well.[60]

And though initially it seemed the Pittsburgh model of focusing shops beyond security seemed outdated, as airport officials continued to renovate and add retail space, they continued to locate such space beyond security where it could serve departing and connecting passengers while they waited for their flights. And as further evidence of the renewal of airport retail, by 2003, BAA faced stiff competition in the retail mall development industry as traditional mall developers found once again that revenue per square foot at airports outpaced that at traditional malls.[61]

New terminal facilities and a fresh approach to airport concessions helped local officials respond to the demands of the new jet age and the era of deregulation. There was one problem that the new jet airliners and increased air traffic brought that continued to confound local and federal officials and that was the issue of aircraft noise. A tremendous amount of energy and money was directed at the problem, but early in the twenty-first century, the noise issue remains the most significant problem local airport officials face.

Notes

1. For a discussion of the "selling" of airline travel, see Courtwright, *Sky as Frontier*, 132–150. For a discussion of the condition of early airport terminals, see Gordon, *Naked Airport*, 41–44.

2. Glenn Garrison, "Inadequate Facilities Jeopardize Jet Era," *Aviation Week* 67 (December 9, 1957): 40–41.
3. "Commercial Jet Outlook Bright for U.S. Airways and Airports," *Aviation Week* 68 (March 3, 1958): 135.
4. Ibid., 135–136.
5. Glenn Garrison, "First Jets Will Use Interim Facilities," *Aviation Week* 68 (March 17, 1958): 28–29.
6. "Aviation: Jets Across the U.S.," *Time Magazine* LXXII (November 17, 1958): 82, 87.
7. The assertions and predictions made in 1959 both overestimated and underestimated current and future traffic. In 1958, the number of revenue passenger enplanements was closer to 53 million. The passenger numbers by 1960 reached 62 million. However, by 1970, the number had reached nearly 170 million, far exceeding the predicted number of 118 million. Historical Air Passenger Statistics, Annual 1954–1980, http://www.rita.dot.gov/bts/sites/rita.dot.gov.bts/files/subject_areas/airline_information/air_carrier_traffic_statistics/airtraffic/annual/1954_1980.html (accessed December 30, 2014).
8. Ford Eastman, "Needs of Airports for Jets Stir Debate," *Aviation Week* 70 (March 9, 1959): 155, 157.
9. "Airport Cities: Gateways to the Jet Age," *Time Magazine* (August 15, 1960): 68.
10. Ibid., 77.
11. Braden and Hagen, *Dream Takes Flight*, 112–114, 125–126; Gordon, 217–218. For additional information on the difficulty in predicting future airport and air traffic growth, see Ian Kincaid, Michael Tretheway, Stephanie Gros and David Lews, *Addressing Uncertainty about Future Airport Activity Levels in Airport Decision Making*, (Washington D.C.: Transportation Research Board, Airport Cooperative Research Program Report 76, 2012).
12. For an overview of the airport expansion programs in these cities, see Robert H. Cook, "San Francisco Airport Expansion Started," *Aviation Week & Space Technology* 78 (June 24, 1963): 41, 43; _____., "O'Hare Walking Distance Stirs Criticism," *Aviation Week & Space Technology* 79 (July 15, 1963): 45, 47; _____., "Idlewild Sprawl Poses Transfer Problems," *Aviation Week & Space Technology* 79 (July 29, 1963): 33–34, 37; _____., "Atlanta, Miami Share Concepts, Problems," *Aviation Week & Space Technology* 79 (August 5, 1963): 47, 49.

13. For a sense of the on-going capacity issue in the 1960s through the early 1970s, see Harold D. Watkins, "Airport congestion is Forcing New Wave of Expansion," *Aviation Week & Space Technology* 83 (October 25, 1965): 174–183; "Traffic Growth Swamps Airport Facilities," *Aviation Week & Space Technology* 85 (October 31, 1966): 145–159; Harold D. Watkins, "Traffic Sparks Airport Needs," *Aviation Week & Space Technology* 98 (May 28, 1973): 66–74.
14. "Fiscal, Social Obstacles Slow Airport Advances," *Aviation Week & Space Technology* 95 (November 15, 1971): 35–36.
15. Carole A. Shifrin, "Official Hope Capacity Crisis Will Spur Expansion of Airports," *Aviation Week & Space Technology* 127 (November 8, 1987): 83, 87, 91.
16. Carole A Shifrin, "Deregulation Bringing Airports More Interest in Own Destiny," *Aviation Week & Space Technology* 121 (November 12, 1984): 174–175.
17. "Airport Renovation, Expansion to Increase in 1985," *Aviation Week & Space Technology* 121 (November 12, 1984): 174–175.
18. Hugh Pearman, *Airports: A Century of Architecture* (New York: Harry N. Abrams, Inc., 2004), 48–70; Wolfgang Voigt, "From the Hippodrome to the Aerodrome, from the Air Station to the Terminal: European Airport, 1909–1945," in John Zukowski, ed., *Building for Air Travel: Architecture and Design for Commercial Aviation* (Munich and New York: Prestel-Verlag; Chicago: The Art Institute of Chicago, 1996), 27–49.
19. Pearman, *Airports: A Century of Architecture*, 78–99. For a sense of how urban thinkers believed aviation would transform the city in the early twentieth century, see Nathalie Roseau, "Reach for the Skies, Aviation and Urban Visions: Paris and New York, c. 1910," *Journal of Transport History* 30 (December 2009): 121–140.
20. David Brodherson, "'An Airport in Every City': The History of American Airport Design," in John Zukowski, ed., *Building for Air Travel: Architecture and Design for Commercial Aviation* (Munich and New York: Prestel-Verlag; Chicago: The Art Institute of Chicago, 1996), 76–78.
21. Ibid., 79–81.
22. For more on Minoru Yamasaki and his work (including the World Trade Center and work at the St. Louis as well as other airports, but not Pruitt-Igoe), see Minoru Yamasaki, *A Life in Architecture* (New York: Weatherhill, Inc., 1979).

23. Xerox copy "Buildings for Business and Government" The Museum of Modern Art, exhibition: February 25-April 28, 1957, 30–35, in Raymond A. Tucker Papers, Series One, Box 25, file: Lambert-St. Louis Municipal Airport 10-1-56 to ___, (University Archives); Pearman, 145; Gordon, 177–180.
24. Saarinen would not live to see either of the buildings completed, however. Both opened in 1962, after he died following surgery for a brain tumor in 1961. For information on Saarinen's early life, education, and the start of his career, see Jayne Merket, *Eero Saarinen* (New York and London: Phaidon Press, 2005), 11–53.
25. Brodherson, "'An Airport in Every City,'" 86–88; Merkel, *Eero Saarinen*, 205: Pearman, 140; Gordon, 186–188.
26. "Idlewild to Get New Concept in Terminals," *Aviation Week* 62 (February 28, 1955): 87–88; "Port Authority Defends Idlewild Planning Criticized by N. Y. Mayor," *Aviation Week* 64 (February 6, 1956): 109, 111; Glenn Garrison, "Airline Costs Soar in New Idlewild Unit," *Aviation Week* 68 (June 23, 1958): 33.
27. Susanna Santala, "Airports," in Eeva-Liisa Pelkonen and Donald Albrecht, eds., *Eero Saarinen: Shaping the Future* (New Haven and London: Yale University Press, 2006), 302–303; Merkel, 206–209.
28. "Second D.C. Airport Gets New Backing," *Aviation Week* 63 (August 15, 1955): 38. 41; Walter McQuade, "The Birth of an Airport," *Fortune* LXV (March 1962): 94; Stuart I. Rochester, *Takeoff at Mid-Century*: Federal Civil Aviation Policy in the Eisenhower Years, 1953–1961 (Washington, D.C.: U.S. Department of Transportation, Federal Aviation Administration, 1976), 164–165.
29. Dulles Airport is widely said to the first airport and terminal designed for the jet age. It was the first new airport designed and built from the ground up anticipating the arrival of commercial jet airliners. However, the Los Angeles airport, the design for which was completed by the time Saarinen began his work at Dulles, was also designed with jet airliners in mind. In the case of Los Angeles, it was a case of an existing airport being redesigned for jet aircraft. As with many claims of "first ever"—it depends on how you define it.
30. Santala, 303; Merkel, 219.
31. Santala, 303–304; Merkel, 217–218. A 1996 project doubled the size of the building, adding 300 foot extensions on either end.
32. Santala, 304; Merkel, 216–217, 222.
33. Merkel, 226–228.

34. "Construction starts at new Kennedy Airport terminal," *USA Today*, December 7, 2005; David W. Dunlap, "Renovated T.W.A. Terminal to Reopen as JetBlue Portal," *New York Times*, February 22, 2008; "Final call for JFK's classic TWA terminal." http://www.cnn.com/2015/10/14/aviation/twa-terminal-jfk-airport/index.html
35. "L. A. Airport Expansion Faces Vote," *Aviation Week* 58 (Mar 11, 1953): 104–105.
36. "LA Airport Fights For Expansion Funds," *Aviation Week* 60 (February 8, 1954): 86; "Los Angeles Traffic Outstrips Expansion," *Aviation Week* 63 (December 19, 1965): 95.
37. "Los Angeles Terminal to Expand," *Aviation Week* 67 (November 4, 1957): 45.
38. "Los Angeles Starts Jet Age Terminal," *Aviation Week* 67 (December 2, 1957): 41–42; Pearman, 151.
39. Jennifer Steinhauer, "In Los Angeles, the Saucer Is Ready to Land Again, *New York Times*, April 17, 2010.
40. "Observation Deck at LAX Theme building to Reopen," http://www.scpr.org/news/2010/07/06/observation-deck-lax-theme-building-set-re-open-pu/ (accessed August 5, 2010).
41. Gordon, 217–219; 231–238.
42. Harold D. Watkins, "Airlines' Fiscal Woes Buffet Airports," *Aviation Week & Space Technology* 95 (November 15, 1971): 39–43.
43. "Airports Devise Economic Survival Plans," *Aviation Week & Space Technology* 115 (November 9, 1981): 207-, 209; "Interest Rates, Economy Deter Airport Financing," *Aviation Week & Space Technology* 115 (November 8, 1981): 202–203.
44. Gordon, 248–257.
45. Dempsey, Paul Stephen, Andrew R. Goetz and Joseph S. Szyliowicz, *Denver International Airport: Lessons Learned* (New York: McGraw Hill, 1997), 359–363.
46. Ibid., 360–367.; Gordon, 248, 254.
47. Dempsey, Goetz and Szyliowicz, *Denver International Airport*, 368–390.
48. "Most Beautiful Airports," http://www.travelandleisure.com/articles/worlds-most-beautiful-airports/1 (accessed August 9, 2010).
49. Marc Dierikx and Bram Bouwens, *Building Castles of the Air: Schiphol Amsterdam and the development of airport infrastructure in Europe, 1916–1996* (The Hague: SDU Publishers, 1997), pp. 252–253. The

BAA model was also adopted by airports in Europe, including Schiphol Amsterdam in The Netherlands.

50. Gordon, 140, 162, 166, 184. For early examples of non-aviation revenue producing activities at airports, see Bednarek, *America's* Airports, 79–85.
51. For a discussion of Pittsburgh airport project and its connection to regional economic development initiatives, see Allen Dieterich-Ward, "From Satellite City to Burb of the 'Burgh: Deindustrialization and Community Identity in Steubenville, Ohio," in James J. Connolly, editor, *After the Factory: Reinventing America's Industrial Small Cities* (Lanham, MD: Lexington Books, 2010), 49–85; Christopher P. Fotos, "Pittsburgh Expanding Capacity with Midfield Terminal Project," *Aviation Week & Space Technology* 130 (June 12, 1989): 329, 331.
52. Fotos, "Pittsburgh Midfield," 329, 331; Michael Marriott, "Pittsburgh Builds Airport of Future Now," *New York Times*, November 12, 1991; "Travel Advisory; New Terminal for Pittsburgh," *New York Times*, September 27, 1992.
53. Christopher P. Fotos, "Revolutionary Terminal Opens Era in Pittsburgh," *Aviation Week & Space Technology* 137 (October 5, 1992): 37–38; Edwin McDowell, "For Pittsburgh, a Model Airport at an Immodest Price," *New York Times*, November 8, 1992.
54. Adam Bryant, "Travel Advisory; A Full-Scale Mall at Indianapolis Airport," *New York Times*, November 12, 1995.
55. Neil MacFarquhar, "La Guardia Sees Its Future in Smart, Small Shops," *New York Times*, July 13, 1997.
56. Ibid.
57. Jennifer Steinhauer, "It's a Mall… It's an Airport;… It's Both: The Latest Trend in Terminal," *New York Times*, June 10, 1998.
58. Joe Sharkey, "The Nation; The Airport Wants You To Shop Till You Drop," *New York Times*, July 8, 2001.
59. Matthew L. Wald, "U.S. Airports Becoming No-Buy Zones," *New York Times*, November 18, 2001.
60. Betsy Wade, "Practical Traveler; Making Airports More Tolerable," *New York Times*, May 21, 2000; Kathleen Phalen Tomaselli, "Kids can fly high – inside the airport," *USA Today*, October 20, 2006.
61. Sara Rimer, "Airport Journal; Extra Waiting Time Is Turning Fliers Into Buyers," *New York Times*, February 4, 2002; Joe Sharkey, "Business Travel: On the Road; Shopping at the Airport In Times of Uncertainty," *New York Times*, November 5, 2002; Terry Pristin, "Commercial Real Estate; Retail Experience Now Arriving at the Airport," *New York Times*, July 2, 2004.

PART III

Jets, Noise, and Unhappy Neighbors

Chapter Five: The Broad Problem of Airport Noise: Airports, the Courts, the Federal Government, and the Environment

Shortly after World War II (WWII), a lawyer specializing in airport litigation published a prescient article summarizing several recent court decisions involving lawsuits over airport noise.[1] As he predicted, noise complaints became a major challenge to airport operators and operations during the post-war period. And although airport operators faced a number of environmental issues, none proved as highly visible or persistent as the noise pollution issue. In the many lawsuits resulting from noise complaints, courts refused to designate airports as nuisances, but they increasingly ruled in favor of airport neighbors who complained about noise, particularly after the introduction of jet airliners.

The introduction of regular domestic jet airline service in the USA in December 1958 clearly escalated the noise issue. However, even before that date, the noise associated with aircraft operations met increasing resistance from the people who owned homes near the nation's expanding airports. Similar to the contemporaneous highway revolts,[2] resistance to airport construction and expansion emerged in city after city throughout the second half of the twentieth century. It reached a peak in the late 1960s and early 1970s, when airport neighbors, already incensed about the noise of subsonic jet engines, faced the prospect of putting up with sonic booms associated with the proposed development of an American supersonic transport (SST).[3]

J.R. Bednarek, *Airports, Cities, and the Jet Age*,
DOI 10.1007/978-3-319-31195-1_6

From 1950s onward, federal officials, airline executives, aircraft manufacturers, airport owners and managers, the courts, and residents near the nation's commercial airports waged a long and contentious battle over noise. One central fight involved assigning blame and responsibility for the problem. Intense debates arose over who was responsible for controlling noise in the first place and what level of government should take the responsibility for addressing the issue. A key court decision in 1962 ruled that airport owners and managers would shoulder most of the responsibility for dealing with the noise issue. Eventually, despite generally calling for local action, the federal government assumed some limited responsibility for regulating the noise of aircraft.

A Staggering Burden: Noise, Airport Zoning, and the Courts

Shortly after the establishment of the nation's first airports, neighbors began to complain about noise and other disturbances caused by aircraft. Through WWII, however, courts generally ruled in favor of airport operators. That began to change with an important US Supreme Court decision in *United States v. Causby* in May 1946. The plaintiffs in the case owned land near an airport in Greensboro, North Carolina, which was used by the military during WWII. The plaintiffs had a barn and house located 2220 feet and 2275 feet, respectively, from the end of the runway. The glide path for the military aircraft while on final approach passed only 67 feet above the house. The plaintiffs complained about not only the noise but also the glare from the aircraft landing lights, both of which forced the owners to abandon their chicken-raising business because their frightened birds—as many as six to ten a day—killed themselves as they ran in fright into the walls of their coops. Additionally, the owners themselves suffered from these intrusions, which made them nervous and frightened.[4]

The case reached the Supreme Court where the justices ruled that the military flights had depreciated the value of the plaintiffs' land. In upholding the claim, the Supreme Court held that Congress had asserted "the air [was] a public highway," and, therefore, the flights over the plaintiffs' land did not represent a case of trespass, but the court concluded "if the flight over the property rendered it uninhabitable, there would be a taking compensable under the Fifth Amendment." And that was what the court found had happened. The court then made two distinctions. First, it distinguished this case from a hypothetical one in which a landowner would sue for damages caused by a railroad operating adjacent to plaintiff's land. In such a case, the noise and vibrations caused by the railroad would be "incidental" and created by a "legalized nuisance." In an opinion written by Justice William O. Douglas, the court declared that since the flights operated directly over the plaintiffs' property "the land [was] appropriated as directly and completely as if it were used for the runways themselves."[5]

Second, the court then proceeded to more precisely define navigable airspace and distinguish it from the airspace used by aircraft while over the plaintiffs' property. Federal law defined navigable airspace "as that 'above the minimum safe altitudes for flight.'" The court's decision contrasted that definition with the airspace occupied by an aircraft when flying at the glide angle called for upon taking off and when landing at airports, which was determined by the operational limits of the aircraft. Federal law set the minimum safe altitude for air carriers at 500 feet above ground level (AGL) during the day and 1000 AGL at night. For other aircraft, the minimum safe altitude stood at 300 feet AGL during the day and 1000 feet AGL at night. The airplanes at Greensboro were flying below those minimum safe altitudes when taking off or landing over the plaintiffs' property and, thus, were not technically in navigable airspace. Though the court declared that it would not at that time set a clear limit between navigable and non-navigable airspace, it ruled that in this case the aircraft were not in navigable airspace and, hence, the hardships caused by the noise and glare represented a taking.[6]

The *Causby* decision specifically noted that it involved military aircraft flying from a municipal airport being operated at the time by the military and so ruled that in this case, the military was responsible for the "taking" and liable for the damages sought by the plaintiffs. The military continued to face noise complaints after WWII, particularly after the introduction of military jets. In general, the US military largely decided to deal with the noise issue its jets created by simply acknowledging the "taking" and compensating land owners.[7] The *Causby* decision, however, did not address the issue of responsibility should the aircraft involved be privately owned, such as by a commercial airline, or if a local government operated the airport.

In the event local governments were held responsible, the tools local governments might use to address the noise issue proved rather limited. As the noise issue intensified in the post-war period, federal officials, airlines, and others all called for local governments to use their zoning power to protect airports from incompatible uses, especially residential areas. Specific airport zoning ordinances, however, ran into legal trouble quite early. Early airport zoning laws focused not on protecting nearby property owners from airport activities, but on protecting airport operations from the actions of neighboring property owners. Specifically, these zoning laws limited the height of structures on adjacent property. In one of the earliest cases examining an airport zoning ordinance (*Mutual Chem. Co. of America v. Mayor of Baltimore*, 1939), the court determined that the land-use regulations imposed did not benefit the general public, just the users of aerial transportation, a minority of the population. In 1948, a California court followed that same argument in striking down a Mendocino County zoning ordinance involving height restrictions. It also cited *Causby* in further determining that limiting the use of the airspace above the plaintiff's property qualified as a "taking" and required compensation.[8] In fact, most airport zoning ordinances passed in the 1950s and 1960s focused exclusively on preventing airport hazards through the limitation of the height of structures on adjacent property. And the courts, with one exception, continued to declare such ordinances as invalid. In the mid-1960s, the Florida Supreme Court had ruled in favor of an airport zoning ordinance, determining that it promoted public safety and served the general welfare.[9]

A commentary published in the *Duke Law Journal* in 1965 strongly criticized the arguments courts had used to invalidate airport zoning laws and argued that airport zoning ordinances should be found a valid exercise of the police powers of local governments. The only alternative would be to force local governments to compensate land owners for the "taking" or "damage" caused by the aircraft noise, imposing upon the local governments a "financial burden" that would be potentially "staggering."[10] Airport owners and operators, indeed, feared that potentially staggering burden and argued repeatedly that airport noise was not a land use issue. Rather, it was a technical issue and needed a technical solution.

Aviation's "Pearl Harbor"

On December 16, 1951, a Miami Airlines C-46, circling to return to the Newark airport after fire erupted in its right engine, crashed in nearby Elizabeth, New Jersey, killing all 56 aboard. Less than 6 weeks later, in January 1952, an American Airlines flight hit a six story building less than a half mile from the December crash site, killing 29 people, including 6 on the ground. Needless to say, Newark area residents were greatly alarmed. And then, on February 10, a DC-6 crashed on take-off from the Newark airport, killing 27 people on the airplane and 4 on the ground. Within hours, the Newark airport closed temporarily. New York City residents expressed concerns when the closing of the Newark facility led to increased flights over their communities. They then joined Newark area residents in demanding immediate action on aviation safety when in April a C-46 crashed attempting to land at Idlewild, killing the crew and three people on the ground. Some in New York demanded that LaGuardia and Idlewild be closed as well.[11]

The New York crashes received wide attention, though the public uproar seemingly died down in the following months. However, C.E. Rosendahl, a member of the National Air Transport Coordinating Committee (NATCC), concluded in a paper presented at the Society for Automotive Engineers National Aeronautic Meeting in April 1954 that the fear created in the wake of the Newark-New York City accidents had "served to crystalize ... already existing and mounting resentment against air terminals and their air operations."[12] And in the same year, Fred Glass, the head of the aviation department of the Port Authority of New York and New Jersey (PNYNJ), called that particular series of accidents civil aviation's "Pearl Harbor." He argued that the crashes had caused many in the New York area to "hate every airplane and everything connected with aviation." He concluded that the incidents "threatened the very existence" of the airline industry in the USA.[13] And much of the resentment focused on noise.

Rosendahl also stressed that much of the hostility toward aviation grounded in the noise issue also involved two additional factors. First, though many airports had originally been located in areas remote from cities, the outward growth of cities had resulted in the construction of new housing developments in the vicinity of the existing airports. Second, the number of aircraft operations at these airports was increasing. When all those factors were combined, residents near those airports, who had already complained about noise, now feared that aircraft would crash near their homes and the busier (and noisier) the airport, the greater the fear.[14]

Both the federal government and the airline industry took action in the wake of the New York accidents. On the federal side, in February 1952, President Harry Truman appointed an airport commission. He asked the group, headed by famed aviator James "Jimmy" Doolittle, to investigate the concerns raised and issue a report within 90 days aimed at increasing safety and decreasing noise in the vicinity of airports. The report,

The Airport and Its Neighbors, appeared in May 1952. It acknowledged actions to reduce noise already underway with CAA sponsorship, but held that in the long term noise complaints were unlikely to abate, especially given the anticipated introduction of civilian jet aircraft. The Doolittle Commission presented 25 recommendations for addressing the problems increased activity at the nation's commercial airports created. The recommendations proposed actions at both the federal and the local levels and recognized that the first-generation military jets were already adding to the noise issue and anticipated the problem would only grow once commercial airlines began flying jet aircraft. The report also included extended discussions of the local issues facing airports, the national issues facing airports, and the importance of better airport planning for the future of aviation in the USA.[15]

Specifically, in terms of recommendations aimed at local officials, the commission called for the construction of new airports and the improvement of existing airports. Local governments should be willing to provide a share of the needed funds, but the commission also suggested that federal airport aid funding be made available to communities to both build longer runways and to create secure cleared areas off the ends of those extended runways. Local governments should also make airports part of their master plans in order to ensure quick and easy transportation access to the facilities. And they should act to include enough land within the airport boundaries to allow for runway extensions "at each end *at least* [emphasis in original] one-half mile in length and 1,000 feet wide." To minimize the amount of cleared space an airport would need, the commission recommended that airports be redesigned or initially constructed with single or parallel runways, oriented as much as possible to minimize the areas flight operations surrounding airports affected. In proposing the move away from the then standard multidirectional runway configurations, the committee encouraged renewed research into cross-wind landing gear for aircraft. In addition, the commission proposed using federal airport funds to encourage cities to improve their local community and airport planning, including the passage of more effective zoning regulations.[16] The passage of local land use controls to deal with the aircraft noise problem became the most frequently cited solution in the 1950s and into the 1960s, especially by the federal government and aviation industry officials, if not by local airport officials.

The bulk of the recommendations in the Doolittle report, however, addressed potential actions at the national level. As the *Causby* decision involved the definition of navigable airspace, the commission called on the CAA to clarify its regulations concerning the use of the nation's airspace and more specifically declare what constituted the navigable airspace, particularly in the approach zones surrounding airports. It further recommended giving the CAA the power to certify airports, thus requiring them to meet certain minimum standards in order to remain in operation. To address safety as well as noise, the commission asked the CAA to strengthen the air traffic control rules within heavily congested airspace around airports and to raise the minimum altitude at which aircraft were allowed to circle while holding for landing or otherwise maneuvering in the vicinity of airports. The CAA should also speed up the acquisition and installation of aids to aerial navigation, particularly those designed to control air traffic in the immediate vicinity of airports. And to prepare for new, heavier and faster aircraft, the commission wanted the length of the standard runways for each type of certified commercial airport extended to at least 8500 feet and in some cases to 10,000 feet. Other recommendations included new rules for aircraft ground operations, crew training on noise avoidance, and the shifting of both commercial and military training flights away from large commercial airports.[17]

While the federal government responded to the crashes with the Doolittle Commission, the airline industry answered with the formation of the NATCC. Members included representatives from the airlines, pilot groups, the FAA, aircraft manufacturers, and the PNYNJ. Focused primarily on the problems facing the New York metropolitan region, the NATCC nonetheless recognized that solutions developed in New York could be applied nationwide. The committee worked with airlines to encourage them to take actions that would alleviate the noise problem, such as changing their operational procedures to eliminate as much as possible low flight over residential areas. During VFR (visual flight rules) approaches into the New York airports, for example, airline pilots were instructed to maintain an altitude of 1200 feet AGL prior to landing for as long as possible. Pilots were also encouraged to climb as quickly as possible to 1200 feet AGL after take-off. The NATCC also recommended the use, as much as possible and as conditions allowed, of preferential runways—those runways oriented in such a direction as to minimize operations over residential areas. And the NATCC worked with airlines to move training operations away from the busiest metropolitan areas, which in the case of the New York area reduced take-offs and landings by 25,000 per year. Finally, the NATCC inaugurated a public information campaign with the belief that if the public understood the economic benefits of the airline industry, it would be less fearful and resentful of the noise produced.[18]

Wishful Thinking: Noise and the Future Jet Age

It was clear that years before the first US commercial airline service in late 1958, the federal government, the airlines, and local airport officials all recognized that noise was going to be an issue, not only for existing airports but potentially for any of the new, larger airports they believed the coming jet age and the growth of air traffic in the USA required. But as each searched for possible answers, they generally disagreed on the proposed solutions. Airport officials pushed for technology fixes, but the airlines and the aircraft industry viewed them as expensive and doubted not only their effectiveness, but whether they would be available in time. While airlines and the aircraft industry resisted what they viewed as expensive technical fixes, they championed instead local land use policies and changes to flight procedures, but these, too, had their limits. The FAA pushed for local land use policies and, with the airlines, also hoped that the American public would come to accept aircraft noise as part of modern life. What became clear was that no one answer would prove effective in addressing the noise problem.

Most of the predictions concerning the advent of commercial jet transport acknowledged the severity of the noise issue and the difficulty in finding a solution. In 1953, for example, sound engineer Dr. Howard Hardy warned that there was "little hope of eliminating the noise nuisance by redesigning aircraft, engines or propellers." He noted that current four-engine propeller-driven planes created noise that interfered significantly with the ability to hold a conversation "anywhere within two miles of the flight line." And he stated that "a four-jet transport would double this swath."[19] Though Hardy did not believe that the noise issue could be solved through better aircraft design, others viewed the aircraft manufacturers as key to the solution. In 1954, Fred M. Glass of the PNYNJ stated that the port authority had done all it could to address the noise issue with procedural changes. Working with the airline industry, the PNYNJ had instituted a preferential runway system that directed most flights to runways that would allow arriving and departing airplanes to avoid overflight of heavily populated areas and approach procedures that both kept aircraft as high as possible for as long as possible before landing and had planes climb more quickly after take-off. The airlines, as noted, also moved training flights out of the New York area. What was needed, Glass asserted, was for the aircraft manufacturers to design and build quieter aircraft.[20]

In January 1956, the Airport Operators Council (AOC) issued a statement at a special CAA meeting on the future of jet transportation. AOC members saw great promise in the introduction of civil jet transport, but they also saw a number of problems. While mentioning the expense and other issues involved with lengthening and strengthening runways, the members identified noise as the number one problem facing airports. The airport operators, echoing Glass's the earlier statement, looked to aircraft manufacturers to provide the needed solution. They stated that unless jet aircraft were designed to be less noisy, the public would not accept them. Aircraft needed to be designed not just "to their ultimate physical possibilities," but to operate in the facilities currently available, including airports in heavily congested urban areas.[21]

To begin to test the effectiveness of some of the proposed solutions as well as more clearly measure the scope of the problems, the CAA began a series of important tests of jet aircraft at airports beginning in 1956. By the time of these test flights, using a borrowed USAF B-47 bomber, a 1956 CAA report already had identified "a total of a hundred problems including traffic control, safety, noise, runway limitations, passenger terminal and other airport facilities" associated with the initiation of jet transport operations. However, the CAA was optimistic the tests could reveal solutions for the many problems identified and that the nation's airport operators still had time to make needed changes before jet airline service began.[22]

Concurrently, the PNYNJ hired the engineering consulting firm, Bolt Beranek & Newman, Inc. (BBN), to study the noise issue and offer advice. The firm, known for its work designing an anechoic (non-echoing) test room for General Electric and quieting the operations of a National Advisory Committee for Aeronautics (NACA) wind tunnel near Cleveland, Ohio, offered three possible solutions to the noise issue. First, engine manufacturers could design and build quieter engines. Second, current jet engines could be made quieter through the use of an engine muffler. And third, airlines and airports could "persuade the public to accept the noise." As for the first two solutions, BBN asserted that they might be possible, but not at the present. As noted, Rolls Royce had developed the first turbofan or bypass jet engine, but such engines were still somewhat experimental and US engine manufacturers would not introduce quieter turbofan engines until the early 1960s. And no workable jet muffler even existed. As for the third suggestion, BBN stated that an effective public relations campaign could "increase neighborhood tolerance of airport noise 10 %." The firm also noted that another favorable trend was the increased use of residential air conditioning. People with air conditioning tended to keep their windows closed during the summer and that reduced noise within the house.[23]

As the expected date for the inauguration of scheduled civil jet transport neared, despite the cautions raised by the BBN report, industry officials put more emphasis on the potential development of sound suppression technology. In May 1957, the Air Transport Association (ATA) expressed confidence that jet noise could be "diverted or reduced to a point where it [would] be acceptable at municipal airports and adjacent areas" by early 1959. The airline officials pointed to noise suppression equipment on the aircraft as well as new procedures and techniques for engine run-ups and taxiing. Much of the noise generated by jets while on the ground could be mitigated through the use of barrier fences or rules that would mandate the pilot orient the aircraft during its engine run ups so as to direct the engine exhaust toward other types of barriers or away from adjacent populated areas. Although the airlines expressed a certain amount of optimism concerning noise suppressors, they pointed out that the existing technology came at the cost of engine efficiency and extra weight as well as the price of the refit. They noted a need to find a "compromise ... between the cost, size, and weight and inconvenience and an acceptable noise level in the particular area involved." Yet, perhaps most importantly, the airlines also acknowledged that there still was no solid information on what noise level would be acceptable. That information was needed so that the airlines and aircraft manufacturers might find the balance between the cost to lower the noise and just how much the noise had to be lowered.[24]

At the same time, the CAA tried to emphasize the degree to which the airlines and the airports were ready to inaugurate service while also continuing to voice concerns over the noise issue. The CAA issued its first report on its studies of the problems associated with the coming jet age in the summer of 1956. As noted, that report generally identified what the CAA termed "the 100 problems" facing jet transport operations. A second report came out in August 1957. The latter report emphasized the progress that had been made in tackling those 100 problems. The CAA declared that many—such as foreign object damage, runway length and strength, and fuel reserves—were not as significant as feared. The CAA even suggested that the noise issue was not quite a severe as first thought. As of 1957, most all of the complaints about jet noise involved military jets. Many military jets had engines using noisy afterburners and were sometimes flying fast enough to create sonic booms. The CAA pointed out that civilian jet airliners did not have engines with afterburners and for the foreseeable future would be traveling at subsonic speeds only. And it expressed confidence that noise suppressors would significantly lower jet engine noise.[25]

The confidence in the effectiveness of noise suppressors, though, was apparently gone by the following summer. In July 1958, CAA Administrator James Pyle told a Congressional sub-committee that the industry had not yet developed a noise suppressor that did not significantly decrease engine efficiency. He further stated the aircraft industry had been overly optimistic in its predications of how well the noise suppressor would work in actually decreasing engine noise. Pyle did anticipate that new engine technology—developed "within the next five years"—would help, but current technology simply did not offer a solution. Further, initial tests of the Boeing 707 at New York area airports had led to demands from the PNYNJ that airlines restrict the weight of their aircraft to 190,000 lbs., which was well below the maximum gross load operation of that aircraft.[26]

All the arguments and concerns were, in a way, theoretical until airlines actually began operating jet airliners on regular schedule in the USA. Pan Am began international Boeing 707 jet service in October 1958. National Airlines began limited domestic jet service in December 1958. Then American Airlines initiated regular transcontinental service between Los Angeles and New York in January 1959. Over the following months, other airlines would follow. By the end of the year, the nation's airlines were flying not only Boeing 707s but Douglas DC-8 jet airliners as well. Only a few airports, though, had jet service in 1959. It was at those few airports that airport operators, airlines, aircraft manufacturers, and the federal government first discovered just how big the jet aircraft noise issue really was.

Sounds of the Twentieth Century

Operating a few jets on test flights was one thing. Both airport and airline officials found that operating jet airliners with passengers and on a regular schedule was an entirely different matter. And the first year of operations produced a number of surprises. For example, as American Airline executive F.W. Kolk reported, actual jet airline operations led to the finding that most noise complaints involved landing rather than take-off operations. The FAA and the airlines had anticipated that most complaints would happen during the take-off phase and had required airline pilots to climb out as rapidly as possible. Not anticipating problems with the landing phase, the FAA had allowed for longer-duration flight at lower altitudes on approaches. Early experience taught that operational changes would be needed to alter approach as well as departure patterns. Further, while the early noise suppressors used on the new airliners helped with low-frequency noises, they generally worked to increase the higher-frequency noises, which proved more annoying to residents.[27]

The actual experience of jet airline operations in the USA sparked a new round of debate between the federal government, the airlines, aircraft manufacturers, and local airport owners and managers over how to deal with the noise issue. The newly established FAA proved quite reluctant to take significant responsibility for solving the noise issue. It generally looked to the airlines and, especially, the airport owners to find solutions. In the wake of the first year of jet airliner operations, in September 1960, the FAA issued a report on noise abatement that acknowledged the introduction of jet airliners had led to "a substantial increase in the number of complaints from persons in the airport environment as well as adverse community reaction to turbojet noise at a number of major airports."[28] The agency recognized that the jet noise issue had the potential to impede air commerce in the USA and looked into how all parties could work together to address this major problem.

In terms of federal government action, the FAA recommended that the Federal Housing Administration, the Department of Defense and the Veterans Administration all have access to noise research done at public airports. They could use that information when making decisions on granting mortgage loans for housing built in areas subject to aircraft noise problems. That same type of information could also go to the Department of Health, Education, and Welfare so that it might "discourage the construction of schools and hospitals falling within the defined areas." The FAA itself would use its rule-making power to enforce the use of preferential runways and anti-noise traffic patterns, take noise data into consideration when reviewing applications for grants under the airport aid program, and conduct further research into solutions to the noise problem.[29]

Most importantly, the FAA report called on local airport authorities "to buy or obtain additional land and avigation easements including the right to cause noise." It asked local governments and planning agencies to use their zoning powers "to the extent legally possible to prevent housing development and places of public assembly from encroachment on the airport, particularly in the approach and departure areas of dominant runways." Where local legislation was weak, the FAA encouraged the enactment of laws to strengthen local zoning powers.[30] As noted, such use of zoning authority proved problematic as courts generally struck down such measures as unconstitutional, arguing that they represented "taking of private property without just compensation."[31]

The FAA's call for the use of local land use planning tools as a primary way to deal with the noise issue became a key component in the debate over how jet aircraft noise should be handled and who should bear the cost. It especially highlighted the tensions between the airline and aircraft industries, on the one hand, and the owners and operators of the nation's airports on the other. In 1960, the Airline Pilots Association (ALPA), the ATA and the Aerospace Industries Association (AIA) formed the National Aircraft Noise Abatement Council (NANAC). Conspicuously absent in the initial membership was the AOC. Though invited to join the new organization, the AOC declined because there seemed to be a significant difference between NANAC proposed noise measures and those the AOC supported. For example, the AOC board believed "ATA President Tipton and the airlines have placed much more emphasis on 'selling noise to the community' rather than finding ways of preventing the community from becoming aroused in the first place."[32]

NANAC's policy of "selling noise" in many ways reflected the position held by the FAA, also a member of the NANAC. In 1962, FAA Administrator Najeeb Halaby responded to a letter from Congressman Joseph P. Addabbo (D-NY), who wrote him after watching a CBS documentary on the aircraft noise problem in communities near Idlewild. Addabbo applauded Halaby's statement that something needed to be done about the noise problem, but was deeply concerned that Halaby had said that aircraft noise was simply something that people would have to learn to live with. In his response to the congressman, Halaby replied: "Frankly, and in all candor, we feel the statement that people must live with noise is realistic in the light of our growing society using complex means of transportation and communication. Noise in urban America is one of the prices paid for progress and economic growth."[33]

The previous year, the FAA had issued a publication on aircraft noise, *Sounds of the Twentieth Century*. The pamphlet aimed at educating the public on how and why jet aircraft produce noise with the idea that if people understood the technology better, the noises would not be as startling. Knowledge, thus, would remove the fear and "emotion" from the noise issue. Moreover, the pamphlet emphasized the many benefits brought by the air transportation system in the USA, including the economic benefits to communities with busy airports. Just as industries had once been attracted to a location due "to the availability of water or electric power," now industries were drawn to communities with available air transport, the pamphlet asserted. "The interests of the airport and the community, therefore, are mutual. A busy airport stimulates business and creates a demand for services and commodities that did not exist before. It creates jobs for many hundreds of people living in the nearby communities." And during the tense early years of the Cold War, the pamphlet linked the nation's airports to the nation's defense, emphasizing their potential role in military plans. "The sound of our own aircraft engines can be as reassuring to us as they were to the people of West Germany during the Berlin Airlift."[34]

Rejecting what emerged as the NANAC and FAA positions, at the AOC membership meeting in May 1960, the organization adopted a resolution outlining a "set of principles" it believed necessary to resolving the issues associated with jet aircraft noise. The resolutions called for aircraft manufacturers, the airlines, and the federal government to all take a stronger, more central role in dealing with the noise problem. Aircraft manufacturers were encouraged "to continue their efforts to devise better and more efficient devices to suppress or minimize the noise of aircraft engines." The airlines were asked to utilize the best suppression devises available and to develop flight procedures that would minimize the exposure to people on the ground to aircraft noise upon take-off or landing. The AOC also called on the FAA to develop rules requiring the use of suppression devices and to deny airport access to airlines flying aircraft without such devices.[35] The AOC further responded to the FAA's noise initiatives at its board meeting in early 1961, strongly objecting to the idea that airport owners should be responsible for purchasing "noise blighted" land. Instead, the FAA should accept responsibility for abating noise and it should institute a program of land acquisition.[36]

While the AOC rejected the FAA land use control and acquisition proposal, the NANAC did not. By early 1962, it was preparing to issue its own handbook dealing with community planning calling for airport zoning and land acquisition. The AOC board then sought a means to counter the NANAC initiative. To that end, the AOC board proposed the creation of a new agency, similar to the Urban Redevelopment Agency, "to buy land in noise sensitive areas and resell it for the purposes which would be compatible with aircraft operations."[37] The AOC drive to place more responsibility for noise on the airlines, aircraft manufacturers and, especially, the federal government intensified in the aftermath of a key Supreme Court decision in March, 1962, *Griggs v. County of Allegheny.*

All Aircraft Noise Is Local

The new Greater Pittsburgh Airport opened in May 1952. The next year, a nearby landowner sued the airport's owner and operator, Allegheny County, Pennsylvania. The property owner argued that the noise from aircraft overflying his house made it impossible for him and his family to continue to live there. As the noise forced the family to move, he argued that the county, by operating the airport, had taken his property and he wanted compensation. A local court agreed that a taking had occurred and determined that the petitioner was due $12,690 in compensation. Allegheny County appealed the decision to the Pennsylvania Supreme Court. That court agreed that a taking had occurred, but decided that Allegheny County was not responsible for the taking. Instead, it opined that the "taker" was either the airlines or the CAA. The case was appealed to the US Supreme Court.

In March 1962, the US Supreme Court overturned the Pennsylvania Supreme Court. Justice William O. Douglas, writing for the majority, argued that although the airport did receive federal funding and had the approval of the CAA, the airport owner made the decisions as to "where the airport would be built, what runways it would need, their direction and length, and what land and navigation easements would be needed." Just as the airport owner was responsible for purchasing land for the runways, it was also responsible for purchasing any "air easements necessary for the operation of the airport." Douglas likened the construction of the airport to the construction of bridges. A county that built a bridge would need to have the easements necessary to provide adequate approaches to the bridge. The court concluded that when the county purchased the land to build the airport "it did not acquire enough" land.[38]

Following the *Griggs* decision, a number of state courts also ruled that airport owners were responsible for aircraft noise. Further, whereas both *Caubsy* and *Griggs* had involved direct overflights of property and a degree of noise that forced the owners to abandon their property, these state courts determined that such extreme conditions were not required for the property owner to receive compensation. First, many state constitutions called for payment to property owners not only when a "taking" had occurred but also when their property had been damaged through state action. Second, a number of state courts ruled that a taking or damage might occur even when a direct overflight was not involved. For example, in Ohio, a state court ruled that a property owner could be compensated even though the overflights were not low enough to force him to leave his home. Further, the Oregon Supreme Court ruled in *Thornburg v. Port of Portland* that a direct overflight was not necessary in determining whether a taking had occurred. Airplanes operating near the property produced noise, the noise was a nuisance and the nuisance constituted a taking. And the Washington State Supreme Court echoed this decision *Martin v. Port of Seattle* (1964).[39]

Fearing an avalanche of lawsuits, the AOC responded aggressively in the wake of the *Griggs* and subsequent decisions. Over the next several years, the AOC fought to have Congress declare the federal government, rather than airport operators, as responsible in the case of noise lawsuits.[40] The AOC also continued its efforts to encourage the FAA to conduct research on the technologies needed to quiet jet aircraft noise. It further asked that the FAA use noise as a criterion in certifying new aircraft. And it lobbied to have Congress allow funds from the federal airport aid program be used to purchase lands surrounding airports as part of a noise abatement program. The AOC also repeatedly criticized the NANAC and its emphasis on land use controls as a key solution to the noise issue. Finally, it called on aircraft and engine manufacturers to develop quieter jet aircraft.[41]

Despite the fears of the AOC, noise lawsuits did not financially overwhelm airport operators in the USA in the immediate aftermath of the *Griggs* decision. Individually the damages initially awarded were relatively modest. However, the number of lawsuits did increase significantly and the airport operators had to defend against them all. And the proliferation of such lawsuits across the country put greater pressure on federal officials to find a national solution to the airport noise problem.[42]

Slow Climb: Toward a Federal Airport Noise Program

Though the issue of jet aircraft noise occasionally attracted significant attention from the FAA and Congress in the late 1950s and early 1960s, it remained largely a local issue. In many ways that changed after 1966 when the director of the White House Office of Science and Technology, Dr. Donald F. Hornig, presented President Lyndon Johnson a report calling for a national program to address the issue of aircraft noise. By that time, as the number of airports around the county experiencing high levels of noise complaints from neighbors expanded, so too did the number of members of Congress with unhappy constituents. Several bills were introduced aimed at increasing the power and authority of the FAA to deal with aircraft noise. Further, the USA was involved in a program to develop a SST and it promised to create new issues about not just engine noise, but regular sonic booms. And the late 1960s witnessed the blossoming of the emerging environmental movement in the USA. While the problems of air and water pollution took center stage, noise was also defined as a form of pollution that needed to be addressed.[43] Though aircraft noise was viewed, at least occasionally, as an urgent national problem, the federal jet aircraft noise abatement program evolved only slowly.

In October 1965, Dr. Donald Hornig hosted a panel on jet aircraft noise at the White House. The meeting included representatives from the FAA, the airlines, National Aeronautics and Space Administration (NASA), the USAF, aircraft manufacturers, airport operators, and ALPA. Meeting behind closed doors, each of the 24 panel members submitted papers detailing their positions on the problem and potential solutions. Hornig presented his report on the meeting the following March. The report made eight basic recommendations. These included the formation of a task force to investigate the noise problem at the three airports experiencing the most intense problems (Kennedy in New York, O'Hare in Chicago, and Los Angeles International); analysis to determine the extent of the problem at 25 major airports by 1975; and research on noise evaluation, noise abatement technologies (especially quieter engines), and noise abatement take-off and landing procedures. The report also called for a federal task force to investigate how federal agencies could aid local governments in promoting more compatible land uses and community development in areas around both airports and military bases. Most importantly, the report concluded that the conflicting interests of many of the parties involved with the jet aircraft noise issue had made it difficult to formulate a coordinated approach to the problem. The entity best situated to develop an effective solution was the federal government.[44]

Unlike many government reports, Hornig's did not merely gather dust on a shelf. Within months, much of the federal research and analysis called for in the report was underway. Under the authority of the White House Office on Science and Technology, representatives from a number of federal agencies, including the FAA, NASA, and the Department of Housing and Urban Development (HUD), had organized three groups—a program committee, a policy and evaluation committee, and a management committee—to tackle the research. The research program was broad ranging involving the development of accepted quantitative noise measurement standards, engine and nacelle design, and new air traffic procedures. When briefing a gathering of representatives from more than 100 domestic and foreign airports, the agencies involved noted that land use controls might offer the best solution to the problem. Aircraft noise, they cautioned, could only be reduced so much. Therefore, the group was examining the degree to which various federal community development programs could be used to prevent incompatible land uses around airports. HUD was looking into the use of insulation to sound-proof homes near airports, drawing inspiration from a program created to address the noise complaints aimed at London's Heathrow Airport. Additionally, the FAA had drafted and submitted to Congress proposed legislation to give it the authority to include noise as one of the standards for the certification of aircraft.[45]

Between 1966 and 1968, various bills were introduced to amend the Federal Aviation Act of 1958 to specifically authorize the FAA to develop a standard to measure aircraft noise and to establish noise control and abatement regulations, including making engine noise part of the criteria for certifying new aircraft. Anticipating the passage of legislation, the FAA began drafting new certification regulations aimed at ensuring that not only would all new aircraft have to be quieter but would also require that existing aircraft either be re-engined or face retirement.[46]

With passage of legislation likely, industry officials directed their criticism toward the proposed regulations that would result. Even before the law passed and was signed in July 1968, industry officials argued that the proposed regulations could cost airlines up to $4 million per aircraft. They also charged that "the FAA's proposals were highly unrealistic in terms of present technology." A quieter "high-bypass ratio turbofan capable of being retrofitted to current jets" was "at least five to six years away." It was also doubted that airline pilots would accept the new flight procedures. Most importantly, forcing US airlines and aircraft manufacturers to comply with the proposed regulations would put them at a competitive disadvantage in the international market.[47]

Battles over the regulations and the standards for certifying large aircraft continued through 1968 and into 1969. By the end of 1968, the aircraft industry, though, saw FAA action as "inevitable." The FAA published the proposed regulations (Federal Aviation Regulations, Part 36 – FAR 36) for public comment in January 1969. The aircraft manufacturers breathed a sigh of relief as the new standards for aircraft certification would only apply to newly designed aircraft (designs seeking certification after December 1, 1969), not aircraft already flying, including first-generation jet airliners like the Boeing 707 and DC-8. It seems the FAA decided to move much more slowly on the issue of what to do about existing aircraft that could not meet the new noise standards as NASA had been conducting some promising research on engine nacelles that could be retrofitted to such aircraft.[48] Initial results indicated that a retrofit would significantly reduce the engine noise of Boeing 707 and Douglas DC-8 aircraft powered by the Pratt & Whitney JT3D engine, but at a cost of almost $1 million per plane. But these were the oldest planes in the commercial airline fleet and were expected to be retired in the near future. Further, the NASA nacelle had not been tested for newer aircraft, such as the Boeing 727 and 737 and the Douglas DC-9. These aircraft used JT8D Pratt & Whitney engines. And while these turbofans were quieter than the older engines, they did not meet the new FAA noise levels.[49]

While the aircraft manufacturers and the airlines were generally pleased with the initial noise regulations, the airport operators were not. The airport operators, in fact, were displeased generally with both the government's and the aviation industry's response to the noise issue. Though the newly renamed Airport Operators Council, International (AOCI), had somewhat cautiously joined the NANAC in 1963, in March 1968, it abruptly withdrew its membership as the NANAC, despite AOCI objections, continued to support land use regulations as the primary solution to airport noise and had opposed retrofitting existing aircraft.[50] Over the next several years, the AOCI would repeatedly push the FAA to issue regulations requiring the retirement or retrofit of all aircraft not meeting the new noise standards as well as federal funds to cover the cost of noise abatement to airports such as court judgments and land acquisition.[51]

Never acting quickly or boldly enough to satisfy the airport owners and operators, the FAA moved very slowly toward rule-making aimed at quieting jet operations and requiring the retrofit or retirement of non-noise-complaint aircraft. In the early 1970s, the FAA and NASA conducted studies on new noise-abating approach and departure procedures. These new procedures, though, met with stiff resistance from airline pilots on safety grounds as they promised to add to the pilot workload during take-off and landing.[52] By the 1970s, however, a number of states, frustrated by the lack of federal action, were passing or threatening to pass their own noise abatement laws.

The FAA did implement two programs designed to help with noise. The first was the "Keep'em High" program in February 1971, which called on air traffic controllers to keep aircraft at higher altitudes in the vicinity of airports. Then it established the new "Get 'em High" departures procedures in August 1972. These procedures were generally accepted by the airlines and did provide some noise relief.[53] More controversially, though, in 1973, the FAA amended its 1969 noise regulation to include new models of aircraft of older design, including the Boeing 707, variants of which remained in production to 1991. Now all large aircraft built after December 1, 1973, regardless of when they were designed, had to meet the new FAR 36 noise standards.[54]

The process of developing noise regulations through the 1970s also involved the creation of different aircraft noise categories. The categories developed by the FAA and DOT generally matched those created by the International Civil Aviation Organization (ICAO), founded in 1947 as a forum for the discussion and negotiation of international aviation agreements. These categories only applied to civil subsonic aircraft and the subsequent noise regulations calling for phased bans of non-compliant aircraft involved only those weighing more than 75,000 lbs. (airliners), not smaller corporate jets. The aircraft categories were based on the aircraft's compliance with noise standards, as defined in FAR 36. Stage I aircraft, the oldest and noisiest jet airlines, were aircraft that did not meet the first noise standards established in 1969. Stage II and, later, Stage III aircraft, in turn, met the Stage II (1969) and Stage III (1977) noise standards. The definition of the stages also included the dates when aircraft were certified and constructed.[55]

The passage of environmental legislation and the creation of the Environmental Protection Agency (EPA) in the early 1970s added another component of federal action to the efforts to address the jet aircraft noise issue. The Clean Air Act (1970) defined noise as a pollution and it gave the EPA authority to research the effects of noise on human health and welfare.[56] Subsequently, Congress passed the Noise Control Act of 1972 that pushed both the FAA and the EPA to take action on aircraft noise. The act empowered the EPA to recommend aircraft noise regulations to the FAA, which was then required to hold public hearings on the proposed regulations. The EPA launched a major aircraft noise study in late 1972.[57] Over the next two years, the EPA worked to develop aircraft noise regulations. A draft proposal in mid-1973 included a long list of potential procedural and policy changes, including adoption of a two-segment approach procedure then being researched by the FAA and NASA and giving local airport operators more authority to create noise abatement rules, including preferential runways, curfews, and even the banning of certain aircraft types. The proposal also called for regulations that would mandate the rapid retrofit of existing aircraft with quiet nacelle technology.[58] Though opposed by aircraft manufacturers and the airlines, many of the proposals, particularly the nacelle retrofit, found strong support among airport operators.[59]

By early 1974, in what became something of a dual between the FAA and the EPA over who would develop noise regulations, the FAA was drafting its own regulations to require the retrofitting of certain aircraft, powered by JT8D and JT3D engines, with quiet nacelle technology. The proposed regulations called for all airlines (both domestic and foreign) to have at least half their fleet meet FAR 36 noise limits—either through refit or by purchase of quieter aircraft—by June 30, 1976. The entire airline fleet would need to meet the FAR 36 noise level by June 30, 1978. The airlines, both foreign and domestic, opposed the retrofit, especially for JT8D-equipped aircraft as research was underway on a quieter fan for that engine. They did not want to pay the cost of the quiet nacelle ($170,000–$200,000 per aircraft) retrofit only to later have to pay the cost of refanning the engine. Retrofitting the older JT3D-equipped aircraft would cost between $770,000 and $900,000 per aircraft.[60] The retrofit requirement also proved quite controversial in the US Senate where a number of senators pushed for the regulation, while two urged delay. Senator Howard Cannon (D-Nevada), chair of the Senate Commerce Committee's aviation subcommittee, and Senator James B. Pearson (R-Kansas) doubted how much the refit would reduce noise and cautioned against making a rule covering foreign airlines.[61]

Then in January 1975, the EPA announced its own proposed regulations to quiet the noise of jet aircraft that echoed the FAA-proposed regulations, which still had not been acted upon. In announcing its proposal, the EPA noted that "16 million Americans are now subjected to a wide range of aircraft noise" and that "civil lawsuits totaling billions of dollars have been filed by citizens seeking relief against airports in such cities as New York, Chicago, Los Angeles, and Washington, D.C." The proposed EPA regulations pushed for the mandatory retrofit of existing non-noise-compliant aircraft with sound-absorbing engine nacelles. The EPA proposed regulations had the same deadlines as that of the FAA. The EPA proposal, though, extended the retrofit requirement to the nation's fleet of corporate jets.[62] Later that year, the EPA also issued proposed regulations requiring airports to develop noise abatement plans.[63]

Moving from the proposed regulations to actual regulations or any necessary legislation proved a long process, however. The debates were many and varied. Federal officials wanted to make sure that any action taken did not open up the federal government to liability in aircraft noise lawsuits as it continued to insist that airport owners and managers had the sole liability for aircraft noise. However, the FAA also stated that it and Congress shared with state and local governments the responsibility for noise abatement regulations and legislation. And there was debate over whether or how the federal government could aid airlines in paying for any retrofit or replacement program. In early 1976, members of the House Public Works and Transportation Committee, absent FAA action on regulations, introduced legislation to set national aircraft and airport noise abatement standards. The bill included a plan to retrofit existing aircraft to meet noise standards. As part of this bill, Congress considered a plan to set up a fund to help airlines finance costs associated with retrofits or fleet replacement. While Congress debated the bill, DOT Secretary William Coleman gave himself a June 1, 1976 deadline to make a decision on regulations involving the proposed retrofit or phase-out of noisier aircraft. He failed to meet his own deadline.[64]

In response, the State of Illinois threatened to sue the FAA to force it to act on the EPA-proposed noise regulations or show cause why it had failed. The state charged that by not acting the FAA was in violation of not only the Noise Control Act of 1972 but the 1968 amendment to the Federal Aviation Act of 1958 that had explicitly given it the power to regulate aircraft noise. While the Illinois suit focused primarily on the noise issues facing Chicago's O'Hare Airport, forcing action on the proposed regulations would have national consequences. Then in October, the EPA again weighed in with additional proposed regulations for both subsonic and supersonic aircraft. The new regulations would require all aircraft certified after January 1, 1980, meet even an even more stringent noise standards than those currently in place.[65]

And all this was happening during an election year. A final push for action came when President Ford announced at an October campaign stop in New York City that he supported the retirement or retrofit of all aircraft in the US airline fleet not meeting FAR 36 standards (1,550 aircraft at the time) by January 1, 1985. He also called for action to ensure the compliance of foreign aircraft operating in the USA.[66] In response to the presidential address, the DOT and FAA finally issued a policy statement in November 1976. The statement reasserted the idea of shared responsibility for noise abatement, but emphasized that liability still rested with airport owners. It also included a timetable for the retrofit or replacement of non-compliant aircraft. Unlike the earlier FAA and EPA proposals, the November 1976 policy statement suggested a much longer period for airlines to comply, in some ways matching the presidential announcement. Two- and three-engine aircraft were to be in full compliance within six years after January 1, 1977. Four-engine aircraft were to be in compliance within eight years of that same date. It also included a proposal to develop a way to help the airlines finance the cost of the retrofit or replacement, but that would require legislative action. In addition, the policy sought a more cooperative relationship with airport owners and operators in the creation of local noise abatement rules. However, the FAA also warned that it would oppose any rules it deemed as discriminatory or that it viewed as interfering with interstate commerce.[67] That same year airports did receive some help as the reauthorization of the aviation trust fund allowed airports to apply for grants to address noise abatement, including funding for landscaping projects, noise suppression equipment, and the purchase of buffer land.[68]

The policy announcement, though, coincided with the end of the Ford presidency. The Ford Administration did make a strong push to have the funding scheme enacted into law—part of the proposed Aviation Act of 1977—but the law failed to pass.[69] Momentum was in favor of some type of stronger federal action on aircraft noise, however, as it was not just a domestic issue. Many airports across the world, especially in Europe, faced the same problem. As the British and French prepared to start flying the world's first and only supersonic aircraft, Concorde, the issue of aircraft noise—not just engine noise, but the aircraft's sonic boom—produced international action on aircraft noise. The ICAO seriously began debating noise standards in 1968. Any measures agreed upon would become an amendment to the document governing international aviation, the so-called Chicago Convention. The ICAO approved Annex 16 in April 1971. It dealt with the measurement, control, and reduction of aircraft noise in the vicinity of airports. It also established noise standards for aircraft and required its member nations to include them in their aircraft certification process. Over the decade, the ICAO continued to revise and update Annex 16.[70] Therefore, the process of developing a retrofit/replacement policy in the USA happened within the context of rather contentious international negotiations over noise standards.

Debate over funding the retrofit/replacement program for domestic airlines continued to play out during the first two years of the administration of President Jimmy Carter and a number of factors shaped the outcome of the debate. For example, that time period witnessed the passage of the Airline Deregulation Act in October 1978. This move to the free market complicated the argument over federal financial aid to the airlines for a retrofit or replacement program. The airline industry continued to insist that it needed help to cover the costs of compliance. The industry and its supporters envisioned a plan whereby airline passengers would essentially cover the costs through a 2 % tax on domestic airline tickets and a $2 charge on international departures. Each airline would have a special account within the Airport and Aviation and Airway Trust Fund into which the fees would be paid. Airlines could draw on the funds to cover the costs of the retrofits and help with the costs of aircraft replacement. Albert Kahn, the chairman of the CAB and the architect of airline deregulation, argued that the airlines did not need the help. Others felt the government aid should be in the form of loans, not grants. In the end, Congress decided that airlines could fund the retrofits and replacements on their own. Also contested were the effective dates for compliance and, since federal aid was no longer part of the equation, many in Congress pushed for an extension of the deadlines. Finally, there were also some who continued to assert that the answer to the noise problem lay with local action, particularly zoning laws that would keep noise-sensitive land uses, such as housing, away from airports.[71]

More than three years after the DOT and FAA announced its retrofit/replacement regulations, Congress responded with the Aviation Safety and Noise Abatement Act of 1979. The legislation, signed into law in February 1980, established a program to retrofit or retire older, noisier aircraft. However, it also extended the deadline airlines had to either retire some of their two- and three-engine Stage I aircraft or make them compliant with Stage II noise standards. And it applied the noise regulations to foreign aircraft operating in the USA.[72]

Following that start, by the mid-1980s, the FAA began work on an updated noise policy that would call for the retrofit or replacement of all Stage II aircraft. Though by the mid-1980s, the number of Americans subjected to aircraft noise levels deemed unacceptable had fallen to an estimated 5 million (down from the 16 million charged by the EPA[73] 1975), aircraft noise was still a major problem. FAA Administrator Donald Egan, in announcing the inauguration of an FAA initiative to move toward more stringent regulations, argued that while market forces would eventually lead to the shift from Stage II to Stage III aircraft, a deadline was important as a target goal.[74]

Once again the wheels of government turned slowly, but international events helped prompt action in 1989. The Egan announcement came at the same time the European Union threatened to limit the ability of international airlines to continue to operate aircraft not meeting Stage III standards in Europe by not allowing them to register additional Stage II aircraft for use there. This might not only require airlines to more quickly replace Stage II with Stage III aircraft, but could hurt the resale value of Stage II aircraft on the world market. The domestic airlines, however, were opposed to the USA adopting a rapid schedule for the shift from Stage II and Stage III compliance. Some warned that such a schedule would lead to higher ticket prices, cause some struggling airlines to fold, and would limit the entry of new airline companies that typically purchased older aircraft. Despite these misgivings, many industry officials saw new noise regulation as “inevitable.”[75]

Congress responded the following year with the Airport Noise and Capacity Act of 1990, a subtitle of an omnibus budget bill. The legislation called upon the FAA to develop regulations detailing a schedule for the phase out of Stage II aircraft. Congress established December 31, 1999, as the optimal date when all civil subsonic aircraft weighing more than 75,000 pounds must meet Stage III standards in order to operate in the USA. The law did allow for the Secretary of Transportation to grant certain waivers (which was expected), that could delay the deadline to December 31, 2003. The law also limited the number of Stage II aircraft that could be added to the US civil airline fleet.[76] The phase out of Stage II aircraft actually occurred by December 31, 2004.

In the decades after WWII, aircraft noise became an extremely complex problem demanding attention from a number of different parties and interests—airport operators, local governments, federal government agencies and departments, in particular, as well as the airlines and engine manufacturers, to name a few. Thus, the issue of jet aircraft noise highlighted the complex relationship between the federal government and local governments when it came to airports and their operation as part of the national air transportation system.

Notes

1. Henry G. Hotchkiss, "Airports Before the Bench," *Aero Digest* 55 (Aug., 1947): 37, 119–20, 122.
2. For discussions of the highway revolts see Raymond A. Mohl, "Stop the Road: Freeway Revolts in American Cities," *Journal of Urban History* 30 (July 2004): 674–706; ____, "The Interstates and the Cities: The U.S. Department of Transportation and the Freeway Revolt," *The Journal of Policy History* 20 (November 2008): 193–226; William Issel, "'Land Values, Human Values, and the Preservation of the City's Treasured Appearance': Environmentalism, Politics and the San Francisco Freeway Revolt," *Pacific Historical Review* 68 (November 1999): 611–646.
3. A great deal of literature on the noise issue, especially as related to aircraft, appeared in the late 1960s and especially the early 1970s. For examples, see Robert Alex Baron, *The Tyranny of Noise* (New York: St. Martin's Press, 1970); Theodore Berland, *The Fight for Quiet* (Englewood, NJ: Prentice-Hall, Inc., 1970); Clifford Bragdon, *Noise Pollution: The Unquiet Crisis* (Philadelphia: University of Pennsylvania Press, 1970); David M. Lipscomb, *Noise: The Unwanted Sounds* (Chicago: Nelson-Hall Company, 1974); Lucy Kavaler, *Noise: The New Menace* (New York: The John Day Company, 1975).
4. Hotchkiss, "Airports Before the Bench," 37.
5. Ibid.
6. Ibid., 37, 119. For the full court decision, including Justice Hugo Blacks's dissenting opinion, see *United States v. Causby* 328 U.S. 256 (1946).
7. Gordon McKay Stevenson, Jr., *The Politics of Aircraft Noise* (Belmont, CA: Duxbury Press, 1972), 85.
8. "The Validity of Airport Zoning Ordinances," *Duke Law Journal* 1965 (Autumn, 1965): 794–795.
9. Ibid., 794; William F. Baxter and Lillian R. Altree, "Legal Aspects of Airport Noise," *Journal of Law and Economics* 15 (April 1972): 65.

10. "Validity of Airport Zoning Ordinances," 802–804.
11. Wilson, 259–64.
12. C. E. Rosendahl, "Aircraft Noise Problem in Airport Vicinities," *SAE Transactions* 65 (1955): 289.
13. Quoted in "Aircraft Noise: Transport design is key to problem, PNYA says; Official warns of need for better equipment," *Aviation Week* 60 (March 8, 1954): 21.
14. Rosendahl, "Aircraft Noise Problem," 289–90.
15. *The Airport and Its Neighbors: The Report of the President's Airport Commission* (Washington, D.C.: U.S. Government Printing Office, May 16, 1952).
16. *The Airport and Its Neighbors,*16–18.
17. Ibid., 17–21.
18. Rosendahl, 290–94.
19. "Urges Airports Near Noisy Areas," *Aviation Week* 58 (January 12, 1953): 82.
20. "Aircraft Noise," *Aviation Week*, 21–22.
21. Airport Operators Council, "CAA Special Jet Meeting: Statement of Airport Operators Council," *News and Comment: A Weekly Bulletin for Members*, January 11, 1956. [Organizational Archives, Airports Council International-North America, Washington, D.C. – hereafter ACI archives].
22. Glenn Garrison, "CAA Borrows B-47 to Check Jet Problems," *Aviation Week* 64 (April 23, 1956): 145.
23. George L. Christian, "Scientists Tackle Jet Noise Problem," *Aviation Week* 65 (September 3, 1956): 50–56.
24. Ford Eastman, "Airlines Say Jet Noise to Be Cut To Reasonable Level by 1959," *Aviation Week* 66 (May 13, 1957): 47. For more on the costs associated with noise suppression technology in the late 1950s see "Silencing Jet Fleet Will Be Costly Civil Engineers Are Advised," *Aviation Week* 66 (May 27, 1957): 47.
25. "CAA Issues First Study Report On '100 Problems' of Jet Age," *Aviation Week* 65 (July 16, 1956): 27; Ford Eastman, "CAA Says Jet Airliner Problems Ease," *Aviation Week* 67 (August 19, 1957): 37.
26. Robert Cook, "Pyle Says Jet Noise Still Major Problem," *Aviation Week* 69 (July 28, 1958): 30; "Extended 707 Tests Ease Dispute," *Aviation Week* 69 (September 15, 1958): 39.
27. F. W. Kolk, "The First Year of the Jet Age ... Reflections," *SAE Transactions* 68 (1960): 585–86.

28. Airport Planning Branch, Airports Division, Bureau of Facilities and Material, *Aircraft Noise Abatement* (Washington, D.C.: Federal Aviation Agency, Planning Series, Item No. 3, Sept. 2, 1960): 2.
29. Ibid., 8–9.
30. Ibid., 8.
31. Charles S. Rhyne, *Airports and the Law* (Washington, D.C.: National Institute of Municipal Law Officers, 1979), 54.
32. Airport Operators Council, "Minutes of the AOC Membership Thirteenth Annual Meeting, Columbus, Ohio, May 25 and 26, 1960," 14 (ACI archives).
33. Ltr, Congressman Joseph P. Addabbo to Honorable Najeeb Halaby, Administrator, Federal Aviation Agency, January 25, 1962; Ltr. N. E. Halaby, Administrator to Honorable Joseph P. Addabbo, Member of Congress, February 9, 1962 (RG 237, Box 54, File: Noise Abatement).
34. See Federal Aviation Agency, *The Sounds of the Twentieth Century* (Washington, D.C.: Government Printing Office, 1961).
35. "Resolution No. 4: AOC Principle Relating To Aircraft Noise," adopted at the Membership Meeting of the AOC May 25, 1960 (ACI archives).
36. Airport Operators Council, "Minutes, Board of Directors Meeting, January 31 – February 1, 1961," AOC Headquarters, Washington, D.C. (ACI archives).
37. Airport Operators Council, "Minutes, Board of Directors Meeting, January 30 and 31, 1962, Tampa, Florida," 17 (ACI archives).
38. United State Supreme Court, 369 U.S. 84 Griggs v. County of Allegheny, Pennsylvania. For an analysis of *Griggs v. Allegheny County* see Allison Dunham, "Griggs v. Allegheny County in Perspective: Thirty Years of Supreme Court Expropriation Law," *The Supreme Court Review* 1962, 63–106.
39. Baxter and Altree, "Legal Aspects of Airport Noise," 38–40; Stevenson, Jr., *The Politics of Airport Noise*, 60–62.
40. The AOC argument was later echoed in "John M. Werlich and Richard P. Krinsky, "The Aviation Noise Abatement Controversy: Magnificent Laws, Noisy Machines, and the Legal Liability Shuffle," *Loyola of Los Angeles Review* 15 (December 1981): 69–102.
41. See Airport Operators Council, "Minutes of the Business Session of the AOC Fifteenth Annual Membership Meeting and Conference, Honolulu, Hawaii, August 22–23, 1962, 23–26; _____., "Minutes of AOC Board of Directors Special Meeting, Chicago, Illinois, October 8, 1962," 2–7; _____., "Minutes of AOC Board of Directors Regular

Meeting, Washington, D.C., January 23–24, 1963," 19; _____., "Minutes of AOC Board of Directors Regular Meeting, Washington, D.C., April 13–16, 1964," 8–9; _____., "Minutes of AOC Board of Directors Regular Meeting, Wichita, Kansas, July 27–28, 1964,"16–26; _____., "Minutes of Board of Directors Meeting, New York, New York, October 12–13, 1964," 12–13; _____., Minutes of AOC Seventeenth Annual Membership Meeting and Conference, New York, New York, October 14–16, 1964," 16–19; "Resolution No. 2 (1965): AOC Principles – Noise," in _____ Minutes of the Business Session AOC Eighteenth Annual Meeting and Conference, Tampa, Florida, October 25–28, 1965." (ACI Archives).

42. Stevenson, 64.
43. Ibid., 99–109.
44. See Office of Science and Technology, Executive Office of the President, *Alleviation of Jet Aircraft Noise Near Airports: A Report of the Jet Aircraft noise Panel* (Washington, D.C., March 1966); Stevenson, 106; "White House Science Unit Seeks Government Attack on Jet Noise," *Aviation Week & Space Technology* 84 (March 28, 1966): 44.
45. Erwin J. Bulban, "U.S. Mapping Program to Alleviate Noise," *Aviation Week & Space Technology* 85 (October 24, 1966): 45–46.
46. "Noise Plan Threatens Older Jets," *Aviation Week & Space Technology* 88 (March 18, 1968): 324.
47. "FAA Anti-Noise Authority Seen Clearing House Unit," *Aviation Week & Space Technology* 88 (March 25, 1968): 28–29.
48. NASA became involved in quiet jet aircraft engine design research in 1967. The research, conducted at NASA's Lewis Research Center near Cleveland, Ohio, had as its goal "the development of a new turbofan, demonstrator engine that will show a reduction of 15 PNdB (perceived noise decibels) on takeoff and 20 PNdB on landing, compared to current transport engines such as the JT3D manufactured by Pratt & Whitney." The research schedule called for the delivery of the demonstrator engine by FY 1972. NASA and the FAA also had a joint research program aimed at identifying and developing the aircraft design requirements and navigational equipment needed to allow pilots to safely fly new approach paths. And NASA was looking at immediate and longer-ranged changes that could be made to engine nacelles in order to muffle engine noise. Michael L. Yaffee, "NASA Begins Major Engine Noise Project," *Aviation Week & Space Technology* 87 (August 21, 1967): 38–39.

49. See Michael L. Yaffee, "FAA Noise Certification Seen Inevitable," *Aviation Week & Space Technology* 89 (November 25, 1968): 39–50; "New Noise Requirements Detailed," *Aviation Week & Space Technology* 90 (November 17, 1969): 35; Kent, 285–287.
50. "Draft Minutes of AOCI General Board of Directors Meeting, September 10–13, 1967, Boston, Massachusetts," 3–4; "Draft Minutes of the AOCI General Board of Directors Meeting, February 5–6, 1968, San Diego, California," 8–9 (ACI Archives); Stevenson, pp. 119–120.
51. For examples see "Draft Minutes of the AOCI General Board of Directors Meeting, August 19–20, 1968," 2–3; AOCI, "Minutes of the General Membership Business Session, AOCI Twenty-Second Annual conference, October 12–16, 1969, Los Angeles, California,"1–2; "Draft Minutes of AOCI General Board of Directors Meeting, Montreal, Canada, November 8–9, 1970,"16; AOCI, "Minutes of the General Membership Business Session, AOCI Twenty-Third Annual Conference, Montreal Canada, November 10–12, 1970," 25–26; "Draft Minutes of the AOCI General Board of Directors Meeting, Fort Lauderdale, Florida, February 25–26, 1971,"12–13; AOCI, "General Board of Directors Meeting, Lake Ozark, Missouri, July 13–14, 1972,"12; "Draft Minutes of the AOCI General Board of Directors Meeting, Seattle, Washington, June 4–5, 1973," 9; "Resolution No. 7: Noise Abatement Retrofits," adopted at the General Membership Business Meeting of the Airport Operators Council International on October 18, 1973, attached to "Minutes of the General Membership Business Meeting, AOCI Twenty-Sixth Annual Conference, Dallas/Ft. Worth, Texas, October 16 & 18, 1973"; AOCI, "Board of Directors Meeting, Seattle, Washington, June 15, 1975,"10; "Resolution No. 4: Federal Aircraft Noise Abatement Policy," adopted at the General Membership Business Meeting of the Airport Operators Council International on September 23, 1976, attached to "Minutes of the General Membership Business Meeting AOCI Twenty-Ninth Annual Conference, San Francisco, California, September 21 & 23, 1976," (ACI Archives).
52. The FAA and NASA conducted a series of tests on a new noise-abatement approach known as the two-segment approach. The new procedure involved "a 6-deg. upper segment and a conventional 3-deg. lower segment, with transitions between the two at altitudes of 250–400 ft." However, the research found that although the

procedure did result in significant noise reduction, it simply could not be done without an unacceptable increase in the pilot workload. In response, some airlines began working on new approach procedures aimed at keeping aircraft as high as possible for as long as possible by descending at a higher angle. "Reduced-Noise Approach Profiles Tested," *Aviation Week & Space Technology* 95 (September 13, 1971): 38–40; Norma Ashford, H. P. Martin Stanton, Clifton a. Moore, *Airport Operations*, 2nd. Ed. (New York: McGraw-Hill, 1996), 108; William S. Hieronymus, "New Landing Method Aimed at Reduction in Approach Noise," *Aviation Week & Space Technology* 94 (March 1, 1971): 46–47.

53. Kent, 293, 295.
54. Hardaway, 124.
55. Both current and historical versions of FAR 36 are available at http://rgl.faa.gov/Regulatory_and_Guidance_Library/rgFAR.nsf/MainFrame?OpenFrameSet (accessed June 20, 2012). For more on the development of aircraft noise standards at the international level, see David MacKenzie, *ICAO: A History of the International Civil Aviation Organization* (Toronto: University of Toronto Press, 2010), 177–178, 296–298. Small propeller driven aircraft are considered non-stage aircraft and military aircraft are exempt from noise standards.
56. For a contemporary discussion of aircraft noise as an environmental issue, see William B. Becker, "Aircraft Noise and the Airlines," *SAE Transactions* 83, Section 3 (1972); 1936–1940, and Louis H. Mayo, "Consideration of Environmental Noise Effects in Transportation Planning by Governmental Entities," *SAE Transactions* 83, Section 3 (1972): 1941–1954.
57. EPA Press Release, "EPA To Launch Noise Control Program," November 6, 1972 http://www.epa.gov/history/topics/nca/02.html. (accessed July 4, 2011).
58. Charles E. Schneider, "Noise Control Proposals Hit Jets," *Aviation Week & Space Technology* 98 (May 14, 1973): 23–24; _____., "Fragmented Noise Control Sought," *Aviation Week & Space Technology*. 99 (July 9, 1973): 19–20.
59. AOCI, "General Board of Directors Meeting," Denver, Colorado, September 5, 1972, 4–6 (ACI Archives); Erwin J. Bulban, "Airport Officials Hit FAA, DOT on Noise, Fund Issues," *Aviation Week & Space Technology* 99 (October 29, 1973): 34. 39.

60. William A. Shumann, "Key Noise Reduction Decisions Imminent," *Aviation Week & Space Technology* 100 (January 7, 1974): 31–33.
61. William A. Shumann, "FAA Gets Contradictory Noise Guidance," *Aviation Week & Space Technology* 101 (August 12, 1974): 41–42.
62. EPA Press Release, "EPA Proposes Quieting of Jet Airplanes," January 31, 1975 http://www.epa.gov/aboutepa/history/topics/nca/03.html (accessed July 4, 2011). The EPA also proposed regulations for quieting thc nation's fleet of existing propeller driven aircraft.
63. "Airport Noise Proposals Readied," *Aviation Week & Space Technology* 103 (October 27, 1975): 28.
64. "House Panel Readies Own Noise Rules," *Aviation Week & Space Technology* 104 (March 1, 1976): 31; Rosalind K. Ellingsworth, "Noise Abatement Fund Studied," *Aviation Week & Space Technology* 104 (March 29, 1976): 12–13; "Washington Roundup: Noise Efforts," *Aviation Week & Space Technology* 104 (May 31, 1976): 11.
65. Eugene Kozicharow, "Illinois to Sue Over Noise Rules," *Aviation Week & Space Technology* 105 (August 2, 1976): 28–29; Rosalind K. Ellingsworth, "Further Cut in Noise Standards Sought," *Aviation Week & Space Technology* 105 (October 11, 1976): 24–25.
66. Preston, 235–236; "House Panel Readies Own Noise Rules," 31; Ellingsworth, "Noise Abatement Fund Studied," 12–13.
67. See Federal Aviation Administration Noise Abatement Policy, November 18, 1976 http://airportnoiselaw.org/faanap-1.html (accessed July 5, 2011).
68. Preston, 238.
69. Rosalind K. Ellingsworth, "Ford Team Readies Final Reform Try," *Aviation Week & Space Technology* 105 (November 15, 1976): 27–28; Preston, 238.
70. MacKenzie, *ICAO*, 176–178.
71. See "DOT Proposes New Trust Fund To Aid Noise Curb Compliance," *Aviation Week & Space Technology* 106 (January 24, 1977): 30; "Airport Operator Noise Plans, Surcharge on Tickets Proposed," *Aviation Week & Space Technology* 106 (March 14, 1977): 28; "Surcharges Urged for Noise Compliance," *Aviation Week & Space Technology* 106 (May 9, 1977): 32;Rosalind Ellingsworth, "House Unit Eases Noise Rule Deadline," *Aviation Week & Space Technology* 107 (July 4, 1977): 26; "Strong, Varied Opposition Meets Latest Noise Bill," *Aviation Week & Space Technology* 108 (June 5, 1978): 35;

Edward W. Bassett, "Kahn Challenges Aid on Noise," *Aviation Week & Space Technology* 108 (June 19, 1978): 38; "Editorial: The Noise Fraud," *Aviation Week & Space Technology* 109 (August 28, 1978): 7; "Noise Reduction Funding Bill Passes House Easily," *Aviation Week & Space Technology* 109 (September 18, 1978): 38; "Noise Funding Shift to Carriers Urged," *Aviation Week & Space Technology* 110 (February 19, 1979): 35–36; "Delay of Noise Deadline Studied," *Aviation Week & Space Technology* 110 (April 9, 1979): 34; "House Conference Participation Key to Maneuvers on Noise Bill," *Aviation Week & Space Technology* 111 (November 12, 1979): 31.

72. "Congress Votes Aircraft Noise Rule Delays," *Aviation Week & Space Technology* 112 (February 11, 1980): 26.
73. The EPA's role in the aircraft noise debate diminished significantly after Congress defunded its Office of Noise Abatement and Control in 1981.
74. James Ott, "FAA Noise Policy to Stress National System, Fleet Replacement," *Aviation Week & Space Technology* 125 (November 17, 1986): 28.
75. Paul Proctor, "FAA Begins Formal Investigation Of Transport Aircraft Noise," *Aviation Week & Space Technology* 130 (February 6, 1989): 60.
76. Airport Noise and Capacity Act of 1990, Subtitle D of title IX of the Omnibus Budget Reconciliation Act of 1990; Christopher P. Fotos, "New Law to Permit Head Tax, Phase Out Stage 2 Aircraft," *Aviation Week & Space Technology* 133 (November 5, 1990): 31.

Chapter Six: Cities and Jet Noise: On the Ground and in the Air, How to Tame the Planes that Roared

Even before *Griggs* established local airport owners and operators as the responsible parties in complaints, local governments and agencies attempted to deal with the airport noise problem. Perhaps the most prominent to take on the issue was the Port Authority of New York and New Jersey (PNYNJ). Repeatedly, the PNYNJ sought policy and legislative remedies to address noise complaints. Though the PNYNJ put some pressure on the FAA and the airlines to take measures to abate noise, courts, for the most part, struck down the authority's and other local legislative efforts, limiting the ability of local officials to deal with the problem the Supreme Court had placed firmly in their laps. This proved especially true as the PNYNJ sought to ban Concorde, the noisy European supersonic airliner, from its airports.

Though the PNYNJ was an early leader in the noise fight, in many ways leadership of the fight against noise shifted from New York to California in the 1970s. Airport officials in and around the Los Angeles area, especially at Los Angeles International Airport (LAX), looked for ways to respond to noise complaints. Similar to the cases in New York, court decisions in many instances limited or even prohibited local government efforts to address the issue. As a result, Los Angeles airport officials had to turn to land purchase as the ultimate way to redress neighbor complaints.

Though New York and California airports perhaps received the most attention in the national press, noise complaints became a challenge at airports large and small throughout the country. In each case, local

J.R. Bednarek, *Airports, Cities, and the Jet Age*,
DOI 10.1007/978-3-319-31195-1_7

airport authorities and governments had to work within the parameters set by court decisions, federal policies, and local political realities to deal with the noise issue. In each case, a combination of flight procedures, soundproofing, and land purchases formed the core of the response. The land purchases in particular resulted in significant changes to the landscape surrounding the nation's airports.

"My Dear Colleague": The Port Authority Takes the Lead

As the date for the inauguration of commercial jet service in the USA neared, the PNYNJ in many ways took a very aggressive stance regarding jet aircraft noise and the need for airlines and aircraft manufacturers to address it. Austin J. Tobin, the executive director of the port authority, outlined his organization's position in what became known as the "My Dear Colleague" letter, sent to a number of airport operators in Western Europe in July 1958. Tobin asserted that port authority officials and officials from American and Pan Am airlines had met concerning the noise produced by the Boeing 707 and that the airlines had recognized that they may have to accept certain restrictions on their operations to bring noise levels down to what the port authority deemed as tolerable to airport neighbors. He said that the airlines further recognized the failure of Boeing to devise an adequate noise suppressor for the aircraft's engine. The letter went on to state that port authority officials had observed test flights at Boeing's field in Seattle where the company's test pilot successfully flew a fully loaded 707 using a take-off procedure that ensured that the aircraft would first quickly attain a relatively high altitude and then fly on reduced engine power. The procedure did provide some relief, though not enough in the view of the port authority's noise expert. And Tobin doubted that many commercial airline pilots would willingly accept the new take-off procedure or that a fully loaded aircraft operating on a regularly scheduled route could match the take-off performance of the test plane. The letter noted, however, that the port authority and Pan Am had agreed to certain operational restrictions on that airline's series of trial flights scheduled for August 1958.[1]

The "My Dear Colleague" letter provoked strong responses from American and Pan Am as well as Boeing Aircraft. The airlines denied that their officials had concurred with the port authority on the noise problems of the Boeing plane or that they had accepted the idea of operating out of Idlewild under port authority-defined operational restrictions. Boeing issued a statement disagreeing with the comments made about the performance of the plane and the company's noise suppressor. The author of an *Aviation Week* article detailing the conflict suggested that "industry sources" believed Tobin had written the letter as a way of gaining more allies for the port authority's "lonely" stance on jet aircraft noise abatement. And perhaps it did make his stance less lonely as shortly after he sent the letter to the directors of European airports, officials at both the London and Paris airports placed restrictions on jet aircraft operations. As a result, Pan Am trial flights at the London airport observed the same weight restrictions required by the PNYNJ.[2]

At Idlewild, Pan Am flew its initial trial flights under those restrictions, which included a gross take-off weight of 190,000 lbs. (well below the maximum take-off weight of 247,000 lbs.), a commitment upon take-off to reach an altitude of 1500 feet as soon as possible, the scheduling of the trial flights between 8 a.m. and 6 p.m., and the use of only four runways. However, when the PNYNJ extended the trials for another 30-day period, the weight restriction was essentially lifted. Instead, Pan Am was required to "observe the gross weight limits imposed by the Civil Aeronautics Authority on certain runways." Half-way through the extended test period Pan Am had still not taken off from Idlewild with the maximum gross load, but it had exceeded the original 190,000 lbs. restriction.[3]

The PNYNJ did not modify all the other restrictions on jet aircraft operations, though, insisting that the airlines observe those procedures aimed at minimizing noise over nearby communities. For example, when possible, the PNYNJ limited jet aircraft take-offs to runways that had the aircraft climbing out over water rather than populated areas. When winds did not allow the use of the preferred runways, the PNYNJ called on airlines to use take-off procedures that would put the aircraft at an altitude of 1200 feet when over nearby communities. The goal was to reduce the noise level to 112 perceived noise decibels (PNdB) on the ground while the aircraft was flying over these communities. As soon as March 1959, however, the port authority was complaining that not all airlines were meeting its established noise abatement standards.[4]

Noise complaints were already described as "heavy" in volume when the National Air Transport Coordinating Committee (NATCC), founded in the wake of the crashes in the early 1950s, announced its own anti-noise program. The NATCC program, however, differed from that of the port authority's in several ways. First, the NATCC announced that its program would cover all aircraft operating out of Idlewild—new jets and older piston-engine aircraft—whereas the PNYNJ rules had only covered jet aircraft. Second, the NATCC program, though including a priority runway provision, was far less strict in the enforcement of the runway requirement and made the aircraft's gross take-off weight a factor in determining which runway should be used. These procedures, the NATCC argued, would allow airlines to operate their "jets as they were built to be operated and as FAA specifies they shall be from a safety standpoint." After that, the airlines would "then seek to provide the maximum possible noise relief for nearby communities." Not surprisingly, the NATCC program received the approval of all the airlines operating out of Idlewild and as well as the FAA.[5]

The conflict between the port authority and the NATCC brought into stark relief one of the many problems airport and other local officials faced when trying to deal with the noise issue. The NATCC and the FAA argued that only the federal government could issue rules and regulations concerning the flight operations of aircraft in the USA. The port authority, on the other hand, argued that as the operator of the airport, it had the right and power "to enforce its regulations regarding the use of jet aircraft at the airport, and could withdraw permission from a non-conforming airline." John R. Wiley, the Director of Aviation for the PNYNJ, said, however, that such a denial of permission would be "an extreme sanction." Executive Director Tobin concluded that the views of the port authority and the NATCC had become "not compatible" and, thus, decided to withdraw the PNYNJ from membership.[6]

To a certain extent, though, the conflict did force the issue concerning who could regulate aircraft noise. Once the FAA asserted only it had the authority to issue such regulations, pressure intensified for it to take action. And beginning in early 1960 the FAA began the process of establishing noise abatement regulations, based on the NATCC program, for both LAX and Idlewild, the first two airports to experience significant commercial jet traffic. When the proposed rules for Idlewild came up for comment, the port authority immediately criticized them due to significant differences between the proposed regulations and port authority rules. For example, the port authority required that all jet aircraft taking off at night use runways 25 or 22 so that they would execute their climb-out over water. The proposed FAA regulations would allow the use of other runways. The FAA regulations also did not establish a maximum acceptable noise level, whereas the port authority rules did. Airlines, which already voluntarily followed the NATCC program, vowed to follow the new FAA regulations "to the letter." Since only the FAA regulations would carry the force of law, it was felt that the airlines would be under less pressure to voluntarily adhere to any of the stricter PNYNJ rules. Perhaps sensitive to the close relationship between the FAA rules and the NATCC program, the FAA announced that it would withdraw from the NATCC as continued "membership would be incompatible with its rule-making activity in the anti-noise area."[7]

The FAA regulations for Idlewild became effective October 1, 1960. After that point, the PNYNJ rigorously enforced the FAA rules and continued to enforce its own rules concerning night-time take-offs and community noise levels and announced that it would enforce its own rules until the FAA challenged it. The matter seemed to come to a head later that month when the port authority announced it would take Delta Airlines to court for its repeated violations of its rules concerning preferential runways for night-time departures and seek a state court injunction against Delta to force it to follow the noise abatement rules. Delta Airlines, though, sought an out-of-court agreement with the PNYNJ. The position of both actors might have been influenced by the fact that the port authority was currently weighing the decision as to whether or not to allow jets to use its airport in Newark where there was strong opposition to such operations. Delta was the only airline at the time applying to start jet service into that airport. It is possible that the PNYNJ wanted to appear determined to enforce its noise abatement rules in order to lessen local opposition. And Delta wanted to cooperate with the port authority in order to gain approval to operate jets at Newark.[8]

In addition to Delta's eventual cooperation, Pan Am also had a very high degree of compliance with PNYNJ noise abatement goals. The airline's record proved advantageous not only at Idlewild, but also at London Heathrow, where Pan Am, viewed as a proven quiet operator, received permission to depart later than other airlines. Competing airlines felt pressure to match Pan Am's record. TWA, for example, complained that because it had a record of lower compliance it had to take-off from London no later than 11 p.m. If it failed to meet that deadline, the flight would be held overnight. TWA officials then asked its pilots to improve their noise abatement performance, to which the pilots complained that the procedures reduced the margin of safety and threatened to vehemently and publicly protest such requirements.[9]

In fact, many airline pilots became vocal critics of the new flight procedures, which they viewed as requiring them to fly in an unsafe manner.

The largest pilot union, the Airline Pilots Association (ALPA), stated in 1961 that its members would refuse to follow any rules they felt sacrificed the safety of flight. The conflict with the pilots gained high visibility following the March 1, 1962, crash of an American Airlines 707 shortly after take-off from Idlewild, killing all 95 persons on board. While no pilot directly charged that the PNYNJ's insistence on enforcing noise abatement procedures had caused the crash, the day before the accident a pilot from Pan Am, a safety officer in ALPA's local council, wrote a letter asserting that the departure procedures allowed for only a very thin margin for safety.[10]

As pilots continued to complain, the editorial board of *Aviation Week* published an editorial strongly critical of local airport noise abatement rules both in New York and London. Especially controversial was a procedure that had pilots reducing power during climbs over noise-sensitive areas. Not only did the PNYNJ call for it, but the reduced power procedure also had the approval of the FAA and the aerospace industry's noise abatement organization, the NANAC. By June 1962, ALPA, which had originally supported some noise abatement measures, shifted its stance to more closely resemble that of its more militant pilots, especially those in the eastern region, which included New York. By that time, 57 airports across the country had FAA-approved noise abatement rules. These rules, based on the FAA's so-called "Red Book" anti-noise pamphlet issued in 1960, included turns at low altitude, reduced power climbs, higher angles of descent, and preferential runways. The pilots argued that those rules were unsafe. The union also argued that the rules had been drawn up with inadequate ALPA participation. It pointed out that the rules were based on very little actual experience with operating jets as they were based primarily on Pan Am's experience in late 1958, which was also a time when ALPA pilots were boycotting the airline. Hence, only supervisory or management pilots, not line pilots, were involved in the trial flights. The rules were also drawn up at a time when the union was more focused on its fight against the FAA's ultimately successful effort to mandate a retirement age for commercial pilots. The union came out with a statement absolutely rejecting as unsafe a number of the prescribed noise abatement procedures, including low-altitude turns, reduced power departures, and steeper descents.[11]

The pilot revolt came shortly after the Supreme Court issued its decision in *Griggs v. Allegheny County*, which put responsibility for dealing with noise primarily at the local level. Meanwhile, the industry anti-noise organization continued to argue for local action to control land uses around airports in order to minimize noise complaints. Further pressuring local airport operators, other court decisions following *Griggs* placed additional limits on the ability of local governments to address the noise issue either through anti-noise ordinances or zoning regulations.

One such case began on September 5, 1961, when the Presiding Supervisor of the Town of Hempstead, New York, Palmer D. Farrington, sent a letter to FAA Administrator Najeeb E. Halaby, expressing deep disappointment with the FAA's response to noise complaints from local residents. Although the FAA had suggested that it would consider some of the recommendations forwarded by town officials, Farrington wrote that nothing had been done and that the noise situation was getting worse. He prevailed upon Halaby to visit his home while attending the National Aeronautic Association's convention in nearby New York City. In visiting Hempstead, Farrington emphasized that Halaby could become an "eye and ear witness to this unbearable condition." Farrington told Halaby that "emotions among my neighbors have run high," but were "justified" as they saw no end in sight to their plight.[12]

Halaby did not reply to Farrington's invitation. Instead, David D. Thomas, Director, Air Traffic Service, sent Farrington a letter stressing the limits to the FAA's role and responsibility in dealing with aircraft noise. Thomas pointed out that the agency had no specific authority to issue noise regulations. What it could do, however, was issue air traffic rules that, while "to the extent consistent with safety," could "take into account the aircraft noise problem." Thomas also emphasized that the responsibility for dealing with aircraft noise was "mutually shared by many segments of the aviation industry and by the various levels of Federal, state and local governments." The letter then described some of the FAA air traffic rules that addressed aircraft noise problems, while noting that it sought to "foster air commerce and ensure safe aircraft operations" and that it had "to balance the equities of the public airport, the adjacent property owners, the air carriers and the traveling public."[13]

Not surprisingly, given that somewhat unhelpful response, in 1963 the town of Hempstead and ten nearby communities joined together in a safety and noise committee to prepare litigation against airlines, lobby for federal legislation, and to propose and push for the enforcement of local ordinances against jet noise.[14] The Town of Hempstead acted quickly on the last goal, adding a provision to its existing noise ordinance that barred "all aircraft noise that exceeded the level of noise made by a large truck passing at a distance of fifty feet in an open area." The airlines operating out of Idlewild immediately sued, arguing that the enforcement of the ordinance would lead to the closure of the airport. Eventually the PNYNJ, ALPA, and the FAA all joined in the lawsuit on the side of the airlines. The case wound its way through the federal district court between 1964 and 1967 when the court issued its decision striking down the aircraft provision of the ordinance.[15]

The aircraft noise issue gained increased attention with the advent of supersonic airliners, and the PNYNJ also took the lead in the reaction against the use of such aircraft within the USA. During mid-June 1974, Air France conducted Concorde demonstration flights at Boston's Logan Airport, which had emerged as one of the nation's most noise-sensitive major airports.[16] Just the idea of the visit proved controversial as some members of the Massachusetts Port Authority opposed the visit on environmental and economic grounds and challenged the executive director's ability to allow the aircraft to land without board permission. The authority's counsel ruled in favor of the executive director and the visit went ahead. The plane flew at only subsonic speed over US territory, so the sonic boom was not at issue. However, the noise from the plane's very powerful engines posed a potential problem. FAA officials measured the noise upon both take-off and landing at the airport. According to the FAA, the noise was comparable to that of first-generation four-engine aircraft. The next month, the general manager of LAX, Clifton A. Moore, warned the AOCI that it was likely that many US airports would be faced with accommodating the new aircraft.[17]

The issue finally came to a head in 1975 when the British and French petitioned for permission to operate Concorde in the USA at Kennedy International Airport and Dulles International Airport. Fairfax County, Virginia, location of Dulles International, tried to use the court system to block the US government from allowing Concorde to use the airport. However, at the time, Dulles was still both owned and operated by the federal government and the courts ruled in favor of the airport's proprietor. Test flights at Dulles began in May 1976, and the Department of Transportation and FAA promised to issue a noise ruling on supersonic aircraft within 30 days of the end of the trial period.[18] The situation was much more contentious in New York.

In April 1975, the PNYNJ responded to the anticipated British-French request for test flights at Kennedy International with the announcement that it "has no present plans to permit Concorde flights into New York."[19]

Additionally, New York Governor Hugh Carey (D) strongly opposed the idea of Concorde operations at Kennedy. Meanwhile, the FAA held a number of public hearings on the topic in New York and Washington, D.C. It also submitted an Environmental Impact Statement (EIS) in September 1975. More public hearings followed in early 1976. Then on February 4, 1976, Secretary of Transportation William Coleman announced his decision to allow test flights at both Dulles and Kennedy. While the test flights at Dulles began largely on schedule, the port authority and the State of New York launched a protracted battle against the decision.[20]

Citizens and public officials in New York responded in the immediate aftermath of the Coleman decision. On February 22, a group of civic organizations protested at Kennedy Airport. That peaceful demonstration was large, loud, and lawful, but the leader of the coalition, Bryan Levinson, threatened that "lawless acts would be committed to shut down the airport – such as dumping sand on the roads – in future demonstrations." That same month, the New York State legislature hurriedly passed a highly restrictive noise bill that Governor Carey promised to sign. The bill would effectively ban Concorde from using Kennedy as well as require the retrofit of older jets for noise abatement. The bill required amendment of the 1947 law that had authorized the PNYNJ to manage the area's airports, and since the port authority was the product of a bi-state compact, the state of New Jersey would also have to pass similar legislation for the noise restrictions to take effect. New Jersey's legislature did not pass a similar bill.[21]

Despite the failure of the New York state action, on March 11, 1976, the port authority banned Concorde from Kennedy for six months, pending the results of the tests at Dulles. British Airways and Air France filed suit in the federal courts to overturn the decision. The final resolution of the issue involved four court decisions, two by a US District Court judge, Milton Pollack, and two by a US Court of Appeals judge, Irving R. Kaufman. The British Airways-Air France suit first went to the US District Court and Judge Pollack. He ruled in favor of the airlines, arguing that the FAA essentially had ordered the PNYNJ to allow the test flights and the authority could not preempt federal authority by banning Concorde. The US Court of Appeals overturned the decision, on the grounds that noise regulation was a joint enterprise, one exercised as a partnership between local and federal authorities. Therefore, the PNYNJ could ban Concorde so long as the ban was not "discriminatory, arbitrary, and unreasonable." Judge Kaufman then asked the lower court to rule on whether or not the ban fit those criteria. Pollack ruled that the ban was discriminatory, arbitrary, and unreasonable. Agreeing with the lower court, on September 29, 1977, Kaufman upheld the ruling, thus overturning the PNYNJ's ban. Once the Supreme Court refused to hear a further appeal, the way was clear for Concorde to land at Kennedy Airport.[22] Though the PNYNJ continued resisting Concorde flights and pushing for greater noise abatement, by the time it was waging the war against Concorde, in many ways the main battle front against jet aircraft noise generally had shifted to California.

California, Noise Abatement, and the Limits to Shared Responsibility

Air travel expanded tremendously in California after World War II. Much of the growth came in the Los Angeles area where the LAX grew dramatically, especially during the 1950s and 1960s. In those two decades, passenger traffic at that airport swelled by 400 % and 200 %, respectively. In 1937, when the airport was first established, the facility that became LAX included only 623 acres. Expansion during World War II and in its immediate aftermath increased the airport's area to over 2550 acres. By 1967, the airport covered 3300 acres with plans to expand to 3400 acres. And Los Angeles, like New York City, pioneered the jet age. The first transcontinental domestic jet flight took off from LAX bound for Idlewild in 1959.[23]

The growth of the airport coincided with the growth of the city and Southern California generally. Burgeoning residential suburbs collided with expanding air commerce throughout California, but particularly in Los Angeles and other cities in that part of the state. Unhappy residents, determined local governments, and the state government all pushed to find ways to mitigate the environmental consequences of air travel, particularly noise pollution. Local curfews and state noise laws met resistance from the airline industry and most often within the courts. The court cases that arose out of California-based initiatives against jet aircraft noise worked to further define the parameters within which local governments could act.

As one of the first cities to enjoy jet airliner service, Los Angeles, like New York City, faced residents angry over jet aircraft noise for several reasons, including its adverse effect on their property values. A Congressional hearing on the matter in 1960 produced several citizen-generated ideas for fixing the noise problem. Proposed solutions included ending jet operations at LAX, prohibiting take-offs to the east, and transforming LAX into a city park. Airport officials responded by touting actions they had taken to lessen the noise—increasing the instrument glide slope from 2.75 degrees to 3 degrees to keep aircraft higher over populated areas while on approach to the airport, requiring pilots to follow a take-off course that would skirt a nearby beach community and to climb to 1500 feet as rapidly as possible, and the prohibition of night-time departures to the east over populated areas whenever winds allowed. The following year Los Angeles County assessors responded to complaints about property values by lowering the assessed value of homes near the airport by 5 to 20 %, a step that cut assessed values by $1.5 million and produced a tax loss of $120,000.[24]

The noise issued figured prominently when city and airport officials embarked on an ambitious airport expansion program in 1967. By the time the master plan was approved, officials had already initiated a program to purchase noise-affected residential properties bordering the airport to the west and the north. The master plan envisioned the purchase of additional properties, including most of the coastal community of Playa Del Ray. The city and the airport viewed those purchases as necessary for noise abatement purposes as well as runway extension plans. Residents, however, often viewed them as destructive of their communities because airport expansion displaced thousands of people. Angry residents could no longer vote their disapproval, however, as a 1963 charter amendment had freed the airport from seeking voter approval before issuing revenue bonds, so they took their grievances to the courts. By the late 1960s, LAX led the nation in airport litigation.[25]

Those affected by noise found some relief in court. Additionally, other decisions in the 1970s continued to limit local airports in their efforts to develop policies and procedures to abate aircraft noise. While the courts continued to assign sole liability for noise to the local authorities, they asserted federal preemption over most attempts to control airport noise, especially night-time curfews. The first major case involved the city of Burbank and its ordinance establishing an 11 p.m. to 7 a.m. curfew on jet operations. The Burbank case, though, involved a very unusual situation, and it was upon that situation that the decision turned. The commercial airport in Burbank, the Hollywood-Burbank Airport, was at that time the only major commercial airport in the country still under private ownership. It was owned by the Lockheed Corporation. Following the passage of the ordinance, the Lockheed Corporation sued Burbank. The case reached the Supreme Court where the court ruled in *Burbank v. Lockheed Air Terminal, Inc.*, in favor of Lockheed. The court's reasoning was very important. It ruled the curfew impinged on an area—control of air traffic—preempted by the federal government, specifically the FAA and, when involving aircraft noise, the EPA. Further, since the city of Burbank was not the proprietor of the airport, it could not exercise its police powers in this case. The decision, though, had an important footnote—footnote 14. In it the court suggested a so-called "proprietor exemption" existed—that a municipal proprietor of an airport might have more authority to pass noise abatement ordinances than did a non-proprietor municipality.[26]

The "proprietor exemption" and its precise meaning became central in a series of cases that then followed. In 1969, the state of California passed legislation that directed its Department of Aeronautics to establish state aircraft noise standards. The Air Transport Association sued arguing that the airlines should not be subject to state-level noise regulations. In this case, a federal district court decided generally in favor of the state of California. In *Air Transport Association v. Crotti*, the court ruled, citing *Griggs*, that since airport proprietors were liable for the damage caused by aircraft noise, they had the right to establish reasonable regulations. Much the same reasoning in *Crotti* also led a federal court judge to uphold a Hayward, California, ordinance establishing a night-time curfew for aircraft that exceeded a certain noise level. In that case, though, the airport involved was a non-commercial airport and the court did not see adequate evidence to indicate that the curfew disrupted interstate commerce. In cases where curfews might impinge on interstate commerce, the court suggested, they might not be permissible.[27]

Although the *Crotti* decision also seemed to suggest that an airport proprietor had the authority to determine what type of air service it desired, the FAA later opposed efforts by California airports to limit air service. For example, in 1980, the FAA opposed the efforts by the Orange County Airport (later John Wayne Airport) to limit scheduled airline service. Orange County Airport officials had not only limited the expansion of service by existing airlines, but also refused permission for other airlines to begin service at the airport. In 1980, two airlines served the airport, Air California and Hughes Airwest. Together, they utilized all the 40-per-day departures slots allowed. The FAA argued that by not allowing other airlines to serve the airport, and effectively granting the two incumbent airlines all the departures slots, airport officials were granting them exclusive rights, an action forbidden under FAA regulations at airports receiving federal airport aid. This test between the FAA and Orange County came in the early days of deregulation, which was supposed to give airlines greater freedom to both enter and exit service at the nation's airports. Airport officials balked when the FAA proposed they hold an auction for some of the airport departure slots, an action aimed at giving other airlines a chance to gain access while still protecting slots held by the incumbent airlines.[28] Eventually, the county and airport managers developed a plan to expand air service at the Orange County Airport.

While the limits placed on airport proprietors and their ability to enact noise control regulations generally worked against them, there was at least one case where it worked in the airport operator's favor. A group of residents in San Diego, California, sued the San Diego Unified Port District, the owner and operator of that city's commercial airport, over its failure to enact noise abatement regulations, such as a curfew. In *San Diego Unified Port District v. Superior Court*, the California appeals court ruled in favor of the port district stating that since federal preemption would prevent the port district from establishing a curfew at the airport—since in this case it would disrupt interstate commercial—the port district could not be held liable for not taking action it was prevented from taking in the first place.[29]

In all these cases involving local airport noise regulations, though, the principle of sole local liability for noise damage was upheld. These cases also upheld the idea of a shared responsibility for noise abatement, but very tight and narrow parameters were drawn around the areas in which local airport owners could actually act to abate noise. Although arguments were made in favor of shifting liability for aircraft noise to the federal government,[30] the FAA in particular was very cautious in establishing its noise abatement regulations in order to shield the federal government from any liability. From the 1960s onward, one of the few areas of common ground between airport operators and the airlines was their call for a federal noise abatement policy. The airport proprietors wanted such a policy in the hopes that it might shift some of the liabilities for aircraft noise from them to the federal government. The airlines wanted such a policy as they did not want to face a myriad of unique local aircraft noise abatement regulations, particularly airport curfews. The FAA, even though it became more responsible for aircraft noise abatement after the 1970s, carefully shaped and worded its policies so as to maintain the sole liability of airport proprietors and avoid shifting liability to the federal government.[31]

Within the limits placed on them by federal policies and the courts, local airport officials often found that the most effective way to limit exposure to noise lawsuits was by buying out the unhappy neighbors, and that was the case in Los Angeles beginning as early as the 1960s. LAX expanded greatly in that decade due to an extensive program to purchase nearby residential areas, especially north and west of the airport. These actions, though, eventually met community resistance. By the 1990s, airport and city officials agreed not to purchase nearby homes unless they were offered by the residents. They also agreed to limit the areas where any purchases would happen. This attempt to be a better neighbor, however, did little to stunt community opposition to airport expansion.[32]

Once Again, All Airport Noise Is Local

While nationally much of the media attention on jet noise abatement and the legal and constitutional issues involved focused on the nation's largest and busiest airports, particularly those in New York and California, aircraft noise became an issue virtually everywhere airports and suburban and residential development collided. The timing of the conflicts varied from place to place as both the degree of residential development and the intensity of jet aircraft operations varied. Also, the responses tended to be shaped not only by the key legal decisions that placed parameters around local actions, but also by the local political environment. Some communities found limited ways to use land control measures—specifically the creation of industrial parks—to shape land use near their airports. Minneapolis-St. Paul developed an aggressive noise abatement program that included both public outreach and extensive soundproofing of homes. In most other places, local airport operators found their ability to deal with noise further complicated by the fact that their often distant airports were surrounded, at least in part, by other, often hostile, jurisdictions. Such was the case in St. Louis and Cleveland. Perhaps the most complex situation developed as the Greater Cincinnati International Airport expanded during the 1980s and 1990s. That airport is owned by Kenton County, Kentucky, and is located in Boone County, Kentucky, but many of the early noise complaints came from the suburbs of Cincinnati, Ohio. Though soundproofing and land use regulations were common, a majority of airports, large and small, eventually resorted to programs to purchase and demolish nearby non-conforming properties, especially residential areas. That became the ultimate way for local airport authorities to control the land use around their facilities and deal with the aircraft noise issue.

Tucson, Tampa, and Detroit: Noise Control through Industrial Parks

While residential development near airports could hamper future growth, some business activities—such as warehousing and shipping—not only benefited from locations near airports, but also were less likely to generate complaints about airport noise or to oppose airport expansion due to noise issues. In some cases, local civic leaders were able to use land use controls and other measures to promote and shape the type of developments surrounding their airports.

Though cities were limited in the degree to which they could shape land use around their airports, both planned and unplanned development activities beginning in the 1950s worked to create both a more compatible built-environment in the vicinity of some airports and a degree of economic development. Historian Douglas Karsner examined development between the 1940s and 1980s around airports in Tucson, Arizona; Tampa, Florida; and Detroit, Michigan. In each case, commercial and industrial facilities drawn to the post-war airports resulted in more airport-compatible land uses as jet noise proved less a problem for them. By the 1950s, for example, the Tucson airport and its vicinity emerged as a preferred site for new aviation-related industrial activity. The transformation began in 1950 when Grand Central Aircraft Company, with a contract to modify B-29s for use in Korea, occupied hangar space at the airport. In 1951, Hughes Aircraft built a missile plant on airport land. The Tucson Airport Authority had actively recruited these companies, and by the end of the Korean War, members of Tucson's business community, known as the Committee of 100, began work to build on these gains and to further attract light industry to the city. The airport played a crucial role as the companies they sought to attract, such as those in the emerging electronics industry, could use aircraft to quickly ship their small, high-value products. The efforts of the boosters proved successful as over the next decade several electronics firms moved their facilities to the airport or its vicinity.[33]

While some of the development surrounding the Tucson airport came as the result of unplanned activities, much also came as the result of planned activities, especially through the creation of airport industrial parks. These airport industrial parks proved especially significant in drawing commercial and industrial uses compatible with airport operations as "developers planned each industrial park, promoted the area, managed the site, and selected compatible occupants." The developers also used protective covenants and deed/lease restrictions to ensure such features as "architectural design, landscaping, and height restrictions based on federal aviation regulations." Tucson developers used the industrial park model extensively in the metropolitan area. By the mid-1960s, for example, there were five industrial parks in and around the city. One of these was the Tucson International Airport Airpark, which was also the largest. By the mid-1970s, developers had constructed nine additional industrial parks within 4 miles of the airport.[34]

Tampa, Florida, also witnessed a rapid development of the land around its airport beginning in the 1950s. As in Tucson, airport proponents in that city worked to attract commercial and industrial uses compatible with airport operations. In Tampa's case, the construction of the Sky Haven Motel in 1946 represented the first fruits of the effort. During the 1950s, a number of additional commercial enterprises had moved their facilities to the airport vicinity including an additional motel, National Car Rental, and the Hertz maintenance department. Other developments included restaurants and gasoline stations. While a number of small industrial concerns also located near the airport, the types of businesses involved generally reflected the importance of the burgeoning tourist industry in the Tampa economy. These types of development—whether planned and unplanned—served to give the airport vicinity a more business and commercial character, rather than a residential character.[35]

Tampa business boosters also used the device of the industrial park to attract planned development to the airport. The Tampa Chamber of Commerce spearheaded the development of the city's first airport industrial park in 1956. By the end of the decade, four additional industrial parks had been established north of the airport. By the 1970s, the area north of the airport had attracted seven additional industrial parks, making it "the largest single concentration of industrial parks in the Tampa Bay region." The tenants of these industrial parks included "light manufacturers, warehouses, offices, showrooms, and storage facilities."[36]

Though development of the land around the Detroit airport began later than in Tucson and Tampa, nonetheless, beginning in the 1960s the area around the Detroit airport also witnessed the same type of commercial and industrial activities. As in the Tucson and Tampa cases, industrial development parks, such as the Airway Industrial Park and the Metroplex Industrial Commercial and Office Park, attracted such companies as American Hospital Supply Corporation and NAPA, an automotive supply company. These industrial parks also included office buildings and facilities for light industry. Beyond these planned developments, other companies locating in the airport vicinity included restaurants, motels, car rental agencies, air freight forwarders, and the Kerr Manufacturing Company (dental instruments and supplies).[37]

A Metropolitan, Community-Based Approach: Minneapolis-St. Paul, Minnesota

The Metropolitan Airports Commission (MAC) in the Minneapolis-St. Paul metropolitan area first faced noise complaints concerning jet aircraft several years before the advent of commercial jet airliner service. In the 1950s, military flying units from the USAF, Army, Navy, Marines, and Minnesota National Guard, many of which were flying early military jet aircraft, operated out of Wold-Chamberlain International Airport, the area's primary commercial airport. In 1956, for example, 42 % of all aircraft take-offs and landings at the airport involved military aircraft. Similar to what happened in the New York City area, a series of crashes heighten the anxiety and noise sensitivity of airport neighbors. Between 1947 and 1956, six military aircraft crashed at or near the airport resulting in a number of deaths and property damage. MAC plans in the mid-1950s for airport expansion, therefore, met concerted local opposition and a major lawsuit calling for the relocation of the airport. The lawsuit delayed but did not derail expansion plans, however. After protracted negotiations, the MAC acquired additional land for the airport from adjacent Fort Snelling. This came just as the MAC signed an agreement with Northwest Airlines to establish its "main base and headquarters" at the airport. Despite the expansion of the existing airport, proposals to build a new airport occasionally resurfaced.[38]

As noted, during the 1960s and into the early 1970s, the FAA pushed the idea of regional jetports. As noise complaints against Wold-Chamberlain mounted in the 1960s, cresting as more than 400 people attended a Minneapolis city council meeting demanding action, a proposal to construct a new, distant jetport surfaced. In 1970, the MAC proposed building a facility at a site it favored north of the cities at Ham Lake. A less favored secondary site located south of the cities was also identified. The MAC plan, though, had to gain the approval of the Metropolitan Council, a regional planning organization created in 1967 and authorized to review airport plans both under the act which created it and the state's Airport Zoning and Development Act passed in 1969.[39]

The Metropolitan Council twice failed to approve the Ham Lake site. It also declared that its policy would be that if a new airport was constructed, all commercial air traffic would have to move to the new facility and Wold-Chamberlain be abandoned. The airport commission, on the other hand, saw both Wold-Chamberlain and the new airport operating within a system of airports serving the Twin Cities. Eventually, a combination of circumstances led to the general abandonment of efforts to build the second airport. First, as the FAA saw the situation, while Wold-Chamberlain alone would soon be unable to handle the projected increase in traffic, it was likely that those levels would necessitate the operation of both the new airport and the existing one. The idea that Wold-Chamberlain would remain open diminished a good deal of the local support for a new airport. Second, the general economic downturn in the early 1970s made the pressure for either a new airport or expansion of the old one less acute.[40]

In the meantime, the MAC established a Metropolitan Aircraft Sound Abatement Council (MASAC) in 1969, an organization that remained in operation until October 31, 2001. Unlike similar councils elsewhere, the MASAC included community representatives as well as airport users among its membership.[41] The MASAC developed an impressive program to abate aircraft noise, including the use of preferential runways that directed take-offs and landings away from populated areas, an increased glide slope to keep airplanes higher while on approach, voluntary night flying curfews negotiated with the airlines, and the removal of airline training flights from the airport. Altogether, the MASAC, the FAA, the airport, and the airlines developed and adopted a 17-point noise abatement program for Wold-Chamberlain.[42]

The late 1980s and early 1990s witnessed another round of studies examining whether or not the Minneapolis-St. Paul metropolitan area needed a new commercial airport or should the existing airport be expanded. As these studies were undertaken and debated, the MAC developed a noise abatement program under FAR Part 150.[43] Unlike many other airport noise mitigation programs, the MAC program emphasized residential soundproofing over property acquisition. This remained true even after the Minnesota State Legislature decided to support the expansion of Wold-Chamberlain instead of the construction of a new airport. From 1993 through June 2002, the MAC started or completed insulation work on more than 6400 single-family homes in the most noise-affected areas. As of that date only Seattle and San Francisco reported insulating more homes. The plan, however, did not completely eliminate property acquisition as 430 housing units and other properties were acquired and demolished.[44]

The Metropolitan Council also adopted an aviation planning document. Although relations between the MAC and the Metropolitan Council had been sometimes contentious, in response to the state's determination to retain Wold-Chamberlain as the primary airport, the council developed a plan for the development of the airport and its surrounding area, including several policies concerning noise abatement. Based on its statutory authority to review local plans, the Metropolitan Council developed several policy statements emphasizing the need for local comprehensive plans to include zoning that encouraged appropriate and compatible uses within those areas affected by aircraft noise. The Metropolitan Council also called for cooperation between local governments, the airport, and federal and state officials to develop land use plans that would ensure the long-term viability of the airport.[45]

Loss of Communities: St. Louis, Missouri

Commercial jet aircraft were introduced at St. Louis's commercial airport in the early 1960s. By 1963, noise complaints had already reached the city and, working with the FAA, local officials were in the process of purchasing land within half a mile of the ends of its main runways for expansion and clear zones. Noise complaints, though came from as far away as 4.5 miles. As the city's Director of Public Works Walter T. Malloy wrote a disgruntled resident, the city was limited in the actions it could take as the FAA controlled aircraft in flight. He noted that the FAA was directing pilots to fly as high as possible for as long as possible in the vicinity of the airport to help with the noise. However, he concluded the noise problem was "a complex one and not easy to solve." He included in his letter a copy of the FAA's pamphlet on jet aircraft noise, *Sounds of the Twentieth Century.*[46]

The land acquisition program initiated in 1963 continued through the decade, but as air traffic increased at the airport, so too did the noise complaints. By 1970, the situation was becoming acute. Complaints from the nearby community of Berkeley, Missouri, increased following the extension of one of the runways toward part of that community, a residential subdivision known as Doddles Dale. The chairman of the city's airport commission, David E. Leigh, in responding to the increase in complaints, made note of the actions the airport commission had taken to address the noise issue. These included the purchase of over 100 acres in the communities of Bridgeton and Berkeley, the clearance of all houses, and "the transfer back to the communities of this land to be used for public parks." The airport also had policies about night-time engine run-ups. In response to the "serious citizen unrest," a committee had been formed to discuss the noise problem in the City of Berkeley in the hopes of avoiding a major law suit. Following the discussion, the commission voted to acquire the property within the Doodles Dale subdivision, "by purchase, or condemnation if necessary." Approximately 70 homes were purchased and demolished.[47]

Noise complaints did not cease, but for much of the late 1960s and early 1970s, the city of St. Louis, under the leadership of Mayor Alfonso J. Cervantes, pushed for the construction of a new airport to serve the St. Louis region. That airport, to be built in southern Illinois, would effectively solve the noise issue as it would result in the closure of Lambert-St. Louis. However, in 1977, Secretary of Transportation Brock Adams canceled the project. Following that decision, the city of St. Louis then embarked on a major campaign, the first of two between the late 1970s and the late 1990s, to expand the existing airport.[48]

Before beginning the first expansion project, as required by federal law, the city submitted an EIS. The EIS gave rise to a plan to deal with aircraft noise, the St. Louis Airport Environs Plan. Central to the plan were major land purchases in the communities of Berkeley and Kinloch, bordering the south and east ends of the airport, respectively. Although there was hope that some neighborhoods might be preserved through soundproofing, the noise abatement program generally resulted in the clearance of purchased properties. This was especially true in Kinloch.[49]

Kinloch, Missouri, is a historically black community. It began as part of a commuter suburb of St. Louis in 1890. Once property in the southern part of the suburban tract was sold to an African American family, the white families in the area moved out and more African American families moved in. The northern part of the development, on the other hand, remained white. In the late 1930s, in a dispute involving schools, the development split into the independent cities of Berkeley and Kinloch. In 1980, Kinloch had a population of 4455. The St. Louis Environs Plan called for the clearance of much of the southwest section of the city. In exchange, Kinloch officials and the airport commission agreed to cooperate on a development plan for the city known as the "Kinloch Tomorrow Plan."[50]

Soon, however, it became clear that the clearance project would proceed far more quickly than any plan for economic redevelopment of the community. By 1984, the Kinloch Merchant's Association was protesting the effect of the clearance program on businesses in the city. They argued that "acquisition of homes in the City of Kinloch by the Lambert St. Louis International Airport has affected the merchants in Kinloch in the most adverse financial manner, leaving us unable to function economically as minority businesses."[51] The following year the airport did agree to a buy-out of Kinloch businesses adversely affected by the noise abatement purchases, but little was done to realize the goals of the "Kinloch Tomorrow Plan," which had promised "a sound tax base, commercial industrial development and jobs."[52] As the airport began a new round of expansion in the early 1990s, the city and the airport commission once again entered an agreement with the City of Kinloch on the redevelopment of land purchased for noise abatement.[53] However, another two decades passed before a development plan would actually be in place for the land acquired and cleared. In the meantime, the population of the town plummeted. It fell to 449 by 2000 and decreased further to 229 by 2010.

Kinloch was not the only community to suffer from the expansion plans of the airport. Officials in the City of Bridgeton, at the northwest end of the airport, complained to Congressman Robert Young (D-MO) in 1985 that purchases of land by St. Louis for noise abatement were hurting its tax base. Once purchased, the land was no longer on the tax rolls as it belonged to the city of St. Louis.[54] While the expansion and noise abatement program of the 1980s was significant, Bridgeton was much more at the center of the expansion program begun in the 1990s, which involved the construction of a new runway. The two original main runways at the airport were too close together to allow for simultaneous use during poor weather conditions. Also, planes often had to cross the runways, while active, in order to taxi for departure or to approach the terminal. The solution was construction of a second runway north and west of the existing runways to provide the airport with two separate all-weather runways. Officials in the city of Bridgeton reacted to the proposal even before the master plan had been completed. They demanded that St. Louis offer fair and full value to those homeowners who would be forced to sell. They also wanted the city to offer options to homeowners that would allow for the preservation of remaining neighborhoods.[55]

As Lambert-St. Louis embarked on its expansion program in the 1990s, its noise abatement program—described by the mayor as "one of the largest and more effective noise mitigation acquisition programs in the country"—had already purchased over 2000 parcels of land and anticipated purchasing an additional 1200 parcels by 1998 for construction of the new runway, which opened in 2006.[56] The expansion plan anticipated that air traffic at Lambert-St. Louis would increase to 40 million passengers by 2010, up from 20 million in the early 1990s. Instead, passenger traffic grew unevenly in the 1990s and peaked in 2000 at 30,558,991. Following the merger of TWA and American Airlines and the September 11 terrorist attacks, both in 2001, Lambert-St. Louis ceased to operate as a major hub airport and traffic has steadily declined, falling to 12,331,426 in 2010.[57] While the expansion plans had not worked to preserve air traffic in the St. Louis region, hard-hit by the deindustrialization, the land acquisition programs associated with them had dramatically altered the landscape around the airport at an often high cost to nearby communities.

Collateral Damage on an Urban–Suburban Battlefield: Cleveland, Ohio

Cleveland Hopkins International Airport is located on 1900 acres of land at the southwest edge of the city of Cleveland. Over the years, Interstate 480 to the north, a NASA facility to the west, a nature reserve to the south, and a Ford Motor Company engine plant to the east limited room for expansion. The areas north of the airport and, since a 2001 land-swap deal, southwest of the airport are located in the city of Cleveland, but otherwise independent jurisdictions including Brook Park, Berea, Olmstead Falls, North Olmstead, and Fairview Park surround the airport. Those municipalities, especially Brook Park, fought repeated battles with the city over noise and airport expansion plans.

The city of Cleveland first engaged in airport renovations in the mid-1960s. The projects undertaken included a major expansion of and improvements to the terminal, a new parking facility, and the upgrading of roads and sewers. They did not include a new runway or the extension of an existing runway. Among the issues halting runway construction was the contentious relationship between Cleveland and the other jurisdictions surrounding its airport, particularly the city of Brook Park, all of which feared increased noise.[58]

Later proposals and master plans from the early 1970s onward, though, envisioned significant changes to the runways. These plans addressed the fact that as late as 1973 the airport had only one runway of sufficient length to handle jet airliners. The city wanted to build a new parallel runway so that planes could land on one and take-off on the other. Any expansion plan, though, depended upon the purchase and demolition of homes north of the airport in the city of Cleveland to provide for the installation of an instrument landing system and to provide a noise buffer zone. Homes in the path of the landing system (Phase 1 of the acquisition plan) had already been purchased by 1973. Homeowners in the adjacent area (Phase 2) wanted the city to move more quickly to acquire their homes.[59] However, neighboring jurisdictions were fighting the runway construction plan. Numerous political battles over finances and land acquisition played out in the Cleveland city council for years. As a result, action stalled on the runway project throughout the 1970s.[60]

Runway battles continued into the 1980s as airport officials sought to expand capacity by building extensions to the existing runways. Opponents claimed the proposed extensions, especially to the main north-south runway, would expose more areas around the airport to jet aircraft noise. Complaints in areas already subject to jet aircraft noise—both in the city and the nearby suburbs—continued to mount as more and more homeowners pushed the city to expand its land acquisition program. Though a few Cleveland residents lamented the demolition of their neighborhoods and resisted the buy-out, many others pushed for a more extensive and rapid program. As the 1980s came to a close, Cleveland Hopkins still operated with basically one main runway, while city and airport officials once again examined plans to build a parallel jet runway.[61]

A third decade of battles between Cleveland and its suburbs over airport expansion began with the publication in June 1992 of a new master plan for the airport. Far more ambitious than earlier plans, this one envisioned the demolition of over 400 homes, involved taking part of the NASA Lewis (later renamed NASA Glenn) research facility, called for the eventual closing and elimination of a convention and exhibition center, and proposed moving a planned Aerospace Technology Park. The plan—which involved two phases over up to 20 years—carried a price tag of close to $1 billion. Cleveland's mayor, Michael White, was determined to see it happen. Brook Park mayor, Thomas Coyne, was equally determined to fight it.[62] After a decade-long battle between the cities, in March 2001, Cleveland and Brook Park announced a final agreement. It involved a major land swap between the two cities and additionally called for a home acquisition program involving up to 314 residential properties for noise abatement.[63]

While Cleveland and its suburbs fought sustained battles over runway expansion and/or construction plans, nearby neighborhoods suffered from uncertainty. Most of the areas identified for clearance to make way for expanded or new runways were eventually cleared. However, the clearance came much more slowly than many residents, wishing to escape aircraft noise, had hoped. On the other hand, some residents fought to preserve their neighborhoods. In the end, the entire township of Riveredge was depopulated, and it eventually disappeared entirely, annexed by Cleveland and Olmstead Falls. Ironically, though the neighborhood was cleared for an expansion of a north-south runway, that project never came to fruition.

Riveredge township dated to 1926. A small group of residents of Brook Park, following a bitterly contested local election, founded the township, which bordered the Rocky River, Cleveland Municipal Airport, and what became the NASA Lewis Research Center. Between 1926 and the 1950s, a series of annexations by both Brook Park and the city of Cleveland reduced Riveredge township to little more than a 48-acre truck farm. John Baluh purchased the farm—and really the entire township—in 1956 and transformed it into a privately owned trailer park. By the 1960s, the Riveredge trailer park could boast of its own volunteer fire department and police department. More difficult financial times followed, however, and the township declared bankruptcy in 1978.

Five years later, the city of Cleveland proposed to purchase the entire township and clear it to create a noise buffer zone for a planned expansion of the airports' north-south runway. Residents and township's officials protested the plan. They argued that 84 % of the townships 477 residents would prefer to stay. They also threatened to allow nearby Fairview Park to annex the land to prevent Cleveland from purchasing it.[64] Then in late 1983, officials from seven south-western suburbs of Cleveland—including Rocky River, Fairview Park, Westlake, Bay Village, Berea, Lakewood, and Riveredge Township—protested the airport expansion plan. Led by Fairview Park mayor, Richard Anter, II, the group argued that in 1978 then-mayor Dennis Kucinich of Cleveland had agreed to limit air traffic on the existing north-south runway to 5 % of the total. The expansion plan would shift up to 20 % of air traffic to the runway, violating the agreement and subjecting the suburbs to increased aircraft noise.[65] The city, detailing the plans in December 1983, argued that as currently envisioned the expansion of the runways would minimize suburban noise exposure. However, it would require the removal of the trailer park in Riveredge Township as well as the re-routing of Brook Park Road.[66]

As the fight continued into 1984 and 1985, the city moved forward with plans to purchase the township and remove its residents. Fairview Park responded to that action by annexing the land to prevent its use for runway expansion.[67] Cleveland and Fairview Park then engaged in a multi-year court battle over the annexation. The two cities finally agreed to divide the township in 1992. By that time, however, it was in some ways a moot point. The residents had all been moved as the township was virtually uninhabited by 1986.[68] And the new master plan for the airport adopted in the early 1990s abandoned the notion of an extended north-south runway in favor of an exclusive focus on a new parallel runway. Overall, the battle over Riveredge—at least as far as the relocation of residents was concerned—was fairly short-lived—1983–1986. It stood in stark contrast to the multi-decade struggle within West Side neighborhoods of Cleveland.

In 1972, residents of the West Side of Cleveland, particularly the West Park neighborhood just north of the airport, confronted double threats—Interstate 480 cut through the southern edge of the neighborhood and the construction of an Instrument Landing System (ILS) approach for Cleveland Hopkins bisected the area just north of the Interstate. In the fall of 1972, construction of I-480 had created a "dead end" on the southern edge of the neighborhood. The announcement of a two-phase plan to purchase and demolish homes for airport expansion added to the turbulence in the neighborhood. Many of the residents—especially those in the Phase 2 area—felt "trapped" between these two large projects. The residents in the Phase I area had their homes purchased and demolished by 1973. However, the residents in the Phase II area faced far more uncertainty. Two years after the announcement of the airport expansion plan, 150 homeowners still awaited final news on when and if their houses would be purchased. And as battles continued over runway expansion projects, there were doubts whether or when their homes would be needed for additional airport growth. From the beginning, their councilman, George Blaha, pushed for the purchase and demolition of the homes. The city proved reluctant to authorize the funding as federal matching grants were not available due to the uncertainly over the runway plans at the airport.[69] As the runway project stalled, the residents were left in limbo.

In the meantime, those same residents were confronted with aircraft noise. Officials at Hopkins International introduced a number of policies aimed at reducing the noise problem, but complaints mounted during the 1980s.[70] At the same time, residents in nearby suburban communities continued to fight the runway expansion plans that would necessitate the purchase of their homes.[71] In many ways, Cleveland's West End neighborhoods became collateral damage in the runway wars that lasted more than 30 years.

In the mid-1980s, the city of Cleveland announced another two-phase plan to purchase homes in the West End's West Park neighborhood. The plan, announced in 1985, detailed a five-year plan which included the purchase and demolition of approximately 130 homes as well as soundproofing an additional 2150 homes and schools. A year later, homeowners were still waiting to hear when their homes would be purchased. One frustrated homeowner declared: "I hate it here. We tried to sell the house a couple of years ago. But people would come out and hear the airplanes and they'd say 'Forget it!' We don't want to put any more improvements in the house. It's not feasible. We just want to go away." Others complained that the aircraft noise made it impossible to watch TV or talk on the telephone. Some described noise so loud it made their walls shake. Many others in the area designated for soundproofing, though, wanted the buy-out to extend to their streets.[72] One resident interviewed by the local newspaper, however, wanted to stay. He had already moved from Riveredge and "loved" his new home. The retired construction worker said the noise did not bother him.[73]

As the home purchase program moved slowly forward, others in the West Park neighborhood also resisted the demolition plan. The Brewer family—parents and children—owned several homes on Sally Avenue. In 1989, the family contacted local media claiming the city was trying to force them to move out. Their homes were all located in the phase-two part of the noise abatement area. When the city scheduled the demolition of one of the houses on their street, the Brewers threatened to block the bulldozer. The city responded saying that the entire buy-out plan was voluntary—no one would be forced to sell their homes. In the phase-one section, for example, 65 homeowners accepted the purchase offers, but five decided to remain in their homes. At the time of the Brewers' protest, 70 homes were included in phase two of the program and 20 had already accepted offers.[74]

While a majority of homeowners accepted offers—and the city gradually expanded the program to include more homes—even those who moved lamented the loss of the West Park neighborhood. In 1993, the *Plain Dealer* noted: "The West Park community, just north of Cleveland Hopkins International Airport, isn't much of a community anymore." Scattered homes remained, but 178 were gone and an additional 160 houses were scheduled for demolition. The Cleveland city councilman representing the West Side declared: "I think it looks terrible. That's why when people say, 'This is a voluntary program,' I've said, 'Who wants to live on a street where you only have scattered houses?'" Many of the former residents moved to nearby suburbs. They complained of having difficulty rediscovering the sense of community they had enjoyed in their West Park neighborhood. One lamented: "I could probably drop dead in this house and nobody would find me. It has been devastating."[75] In April 1993, however, the city announced plans to purchase and demolish an additional 440 homes.[76]

In 1995, former residents reunited for a reunion. They came back to a neighborhood, though, that was nothing like the one they had left. The noise abatement program, in the words of the local newspaper, had come "close to destroying their old neighborhood." They returned to "an area dotted with vandalized homes, bulldozed lots, half-demolished buildings, weeds and a handful of hardy souls who refused to move away." Only about 30 homes remained in the neighborhood where over 400 had once stood. And as the city continued to explore expanding the clearance area, another 330 homes were in "limbo." The program remained "voluntary," but as one city council member stated: "It's only voluntary, sort of.... I don't personally think it's practical for people to stay on a street forever when houses are torn down. It's voluntary in principle, but as a practical matter, it's not real practical." As the number of remaining residents decreased, complaints about crime and vandalism increased. The airport finally hired private security guards to patrol the area.[77]

Finally, in 1998, the city announced an accelerated program to clear the last 200 or so homes in the areas of the larger West Side area identified as most affected by airport noise. After nearly two decades, both residents and neighborhood institutions felt the strain of the long decline of the area. St. Patrick's Catholic Church in the oldest Catholic parish in the city of Cleveland lost one-third of its parishioners between the early 1970s and 1998. Businesses within the area also suffered with one drug store claiming its sales were down by $1000 per day as the demolition programs accelerated in the 1990s. The city sold most of the area cleared by its noise abatement program to Chelm Properties, which in turn announced a $200 million Cleveland Business Park project aimed at providing initially 2.8 million square feet of office and light-industrial space in its first phase. Yet, some residents continued to resist. Part of the resistance was grounded in the desire to stay where they had lived—some for decades. Most though focused on what they saw as the unfair prices being offered by the city.[78]

Fourteen years later, only phase one of the $200 million business park had been constructed. Much of the land purchased in the 1990s remained vacant. The new parallel runway opened in 2004. Although that part of the expansion program went forward, much of the rest was scaled back or postponed in the wake of the terrorist attacks in September 2001 and the economic recession beginning in 2008, both of which reduced the demand for air travel in the USA.[79] During this time, Cleveland Hopkins has seen a steady downturn trend in the number of passengers served, from 6,269,516 in 2000 to 4,375,822 in 2013.

What Do You Do When the Airport Is in Kentucky and the Noise Is in Ohio: The Greater Cincinnati/Northern Kentucky International Airport

Many examples of airport owners and operators dealing with noise abatement and land use issues involved conflicts across multiple jurisdictions. Perhaps among the most unusual cases involved the jurisdictional complexity surrounding the Greater Cincinnati-Northern Kentucky International Airport. As noted, the airport was owned and operated by Kenton County and located in Boone County, Kentucky. However, when the airport initially expanded in the 1980s as a major hub for Delta Airlines, many of the noise complaints came from jurisdictions in the suburbs of Cincinnati, Ohio. The conclusion that the distant airport board was not hearing their complaints led noise-affected Ohioans to demand a greater voice in the operation of the airport. Further expansion in the 1990s resulted in noise complaints not just from suburban communities in Ohio, but in Kentucky as well.

In the mid-1980s, what is now the Greater Cincinnati-Northern Kentucky International Airport was in the final stages of plans to construct a new north-south runway. The flight path for the new runway, located on the eastern side of the airport, would bring air traffic over Delhi and Green townships west of Cincinnati on the Ohio side of the Ohio River. In October 1986, residents in those townships called for a public hearing on the airport plan in their communities, rather than in Kentucky, where a public hearing had been held in September. Residents, along with their congressman, Thomas Luken (D-Ohio), strongly questioned the decision to build the north-south runway. They favored, instead, either another east-west runway (which would keep the noise largely in Kentucky) or a north-south runway on the west side of the airport (directing the noise to areas further west of Cincinnati). Over 750 residents turned out for the public hearing held in Delhi Township in November 1986 to voice their concerns.[80]

Despite the protests, the Kenton County Airport Board moved forward on construction of the new runway, applying for federal funding in June 1988. Supporters of airport growth on both sides of the Ohio River began a concerted campaign to emphasize the economic benefits of the new runway. The Cincinnati Chamber of Commerce, for example, boasted that the new runway—and the expanded air traffic it would allow—would add up to $1 billion to the local economy by 1992. That campaign, though, further upset opponents who saw it as an attempt to "steamroller public opinion." Following FAA approval of construction funds, the Oak Hills, Ohio, Board of Education, joined Delhi Township in its opposition to the runway, agreeing to join in any litigation. Delhi Township, which had been threatening legal action, finally filed suit in September 1988. And although the township won in the first round, a US district court judge lifted the temporary restraining order, opening the way for the Kenton County Airport Board to move forward on contracts for the construction of the new runway.[81] Meanwhile, the airport board, Delhi Township, and the Oak Hills Board of Education signed an agreement on a noise abatement plan in December 1988. The plan involved flight restrictions and a promise to monitor noise in Delhi Township. Should the noise reach a certain threshold, the airport board agreed to help pay the cost of soundproofing a local Oak Hills school.[82]

Once the runway opened in January 1991, noise complaints began. At this time, Delhi Township residents and other airport opponents decided to challenge the airport board itself, which consisted of seven white males, six appointed by the chief executive-fiscal judge of Kenton County and the seventh by the Boone County executive. The board also consulted with a ten-member advisory board—created in 1964 in the wake of the attempt to build a regional airport in southern Ohio (see chapter "Response to the Jet Age: Federal–Local Interaction and the Shaping of the Aviation Landscape")—with six Ohio representatives all recommended by the judge in Kenton County and appointed by the Kentucky governor. Members of the advisory board could serve on committees and vote in their meetings, but only the seven members of the governing board had final voting authority.

Opposition to the airport board's structure intensified as work began on a new master plan for airport expansion in 1992, including the construction of another north-south runway on the west side of the airport.[83] Congressman Charles Luken (D-Ohio), son of retired Congressman Thomas Luken, wrote the governor of Kentucky voicing his displeasure with the airport board's noise abatement record and asking for Ohio representation. He was later joined by State Representative Jerome F. Luebbers, the Hamilton County Commission, the Cincinnati City Council, Ohio's two US senators, and a newly elected representative, David Mann (D-Ohio), who suggested he would introduce legislation to mandate Ohio representation. This pressure yielded several results. First, the airport board appointed Hamilton County officials as well as representatives from Delhi and Green Townships to serve on the committee looking at the new expansion plan. In 1995, moreover, the airport board allowed for committee decisions to go directly to it for a vote rather than to an executive committee first. However, the master plan had already been essentially approved before the board took any of those mitigating actions. Finally, while the airport board's structure remained the same, change did come to the advisory board in 1998 when the Kentucky governor expanded its membership to 11 and allowed Cincinnati and Hamilton County officials to each appoint one member directly.[84]

While much of the noise fight in the Cincinnati area focused on revising the composition of the airport board, there were also noise complaints on the Kentucky side of the river and these led to land purchases. Kentucky airport neighbors had been willing to tolerate the noise when flights were infrequent. However, as the airport expanded and the number of flights increased, tolerance for the noise decreased. In all, by the early 1990s, more than 2200 residential properties in Ohio and Kentucky occupied areas experiencing significant levels of aircraft noise. And these complaints increased with the opening of the new runway. To make matters worse, a growing number of cargo jets began to use east-west runways, bringing more complaints from Kentucky subdivisions under those flight paths. This led the airport board to begin soundproofing some homes and purchasing others in a number of Kentucky neighborhoods.[85] And in 1999, as the airport prepared to construct a new north-south runway on the west side of the airport, officials announced another round of land purchases and home demolitions, all in Kentucky.[86]

The expansion of the airport certainly forced the board to address noise complaints, but the issue faded somewhat in importance in the early twenty-first century as the airport lost business. Air traffic in and out of Cincinnati, as elsewhere, slowed following the September 11, 2001 attacks. Then, in 2005, ironically shortly after the opening of a new north-south runway, Delta Airlines decreased its flights into and out of its Cincinnati hub. The subsequent merger of Delta and Northwestern in

2008 accelerated the decline in the use of the facility. Between 2003 and 2010, the airport lost two-thirds of its flights. Between 2005 and 2010, a third of the jobs the airport had once generated had gone and the number of passengers using the airport had fallen by half.[87]

When dealing with airport noise issues, local authorities found themselves operating within a very complex local–federal relationship. Both local and federal officials had some responsibility for dealing with the problems associated with aircraft noise and needed to work together, but the shared nature of the responsibility placed strict parameters around local action.

Often the ultimate, and sometimes the only, solution local officials could resort to was the purchase and clearance of land around their airports. Though not as massive as the changes brought by the interstate highway system, nonetheless the advent and expansion of jet airline travel in the USA significantly shaped and changed the landscape around the nation's airports.

Notes

1. "Text of Port Authority's 'My Dear Colleague' Letter,'" *Aviation Week* 69 (September 8, 1958): 40–41.
2. Glenn Garrison, "Hope for Jet Noise Compromise Shaken," *Aviation Week* 69 (September 8, 1958): 39, 41; "Extended 707 Tests Ease Dispute," 39.
3. "Extended 707 Tests Ease Dispute," 39.
4. Glenn Garrison, "Port Authority Noise Investigators Criticize American's Jet Takeoffs," *Aviation Week* 70 (March 30, 1959): 46; "Hot Weather Adds to Jet Noise Problem," *Aviation Week* 70 (June 22, 1959): 100.
5. "Hot Weather Adds to Jet Noise Problem,"100; Glenn Garrison, "NATCC's Idelwild Anti-Noise Plan Offers Economical Jet Operations," *Aviation Week* 70 (June 29. 1959): 29.
6. Quoted in Glenn Garrison, "Port Authority Breaks with NATCC," *Aviation Week* 71 (September 21, 1959): 45; see also "FAA Noise Position," *Aviation Week* 71 (September 21, 1959): 45.
7. Glenn Garrison, "Port Authority, Airlines Analyze FAA Idelwild Jet Anti-Noise Rules," *Aviation Week* 72 (May 16, 1960): 41; "Proposed FAA Noise Rules Challenged," *Aviation Week* 73 (August 29, 1960): 43.

8. Glenn Garrison, "N. Y. Port Authority Takes Delta To Court Over Idlewild Jet Noise," *Aviation Week* 73 (October 31, 1960): 36; "Delta Wants to Resolve Idlewild's Jet-Noise Complaint Out of Court," *Aviation Week* 73 (November 7, 1960): 43.
9. "Jet Crash Spurs Pilot Criticism," *Aviation Week and Space Technology* 76 (March 19, 1962): 42.
10. Ibid.; Stevenson, 26.
11. David A. Hoffman, "Noise Returns as Major Airline Problem," *Aviation Week and Space Technology* 76 (April 16, 1962): 36–40; "Pilots Take Tougher Noise Stand, Define Unacceptable Maneuvers," *Aviation Week and Space Technology* 76 (June 11, 1962): 47; Stevenson, 34–35.
12. Ltr Falmer D. Farrington, Presiding Supervisor, Town of Hempstead to Mr. Najeeb E. Halaby, Federal Aviation Agency, September 5, 1961 (RG 237, Box 30, File: Noise Abatement).
13. Ltr. D.D. Thomas, Director Air Traffic Service to Mr. Palmer D. Farrington, Presiding Supervisor, Town of Hempstead, September 11, 1962 (RG 237, Box 30, File: Noise Abatement).
14. Stevenson, 45–46.
15. Stevenson, 45; "Comment: The Constitutionality of Local Anti-Pollution Ordinances," *Fordham Urban Law Journal* 1 (1972): 213. The court found it was in conflict with FAA regulations as airlines could not comply with the noise ordinance using the FAA-established flight pattern for the airport.
16. For a contemporary examination of the controversies surrounding Boston's airport, see Dorothy Nelkin, *Jetport: The Boston Airport Controversy* (New Brunswick, NJ: Transaction Books, 1974).
17. "U.S. Officials Find Concorde Noise Acceptable," *Aviation Week & Space Technology* 100 (June 24, 1974); 27; "Concorde Stirs Massport Dissension," *Aviation Week & Space Technology* 100 (June 24, 1974); 28; "Airport Chiefs Urged to Prepare For Airline Concorde Operations," *Aviation Week & Space Technology* 200 (July 1, 1974); 25.
18. Stephen L. Schechter, "The Concorde and Port Noise Complaints: The Commerce and Supremacy Clauses Enter the Supersonic Age," *Publius*, 8 The State of American Federalism, 1977 (Winter 1978): 148–149.
19. Quoted in William A. Shumann, "Port Authority Position Clouds Concorde Operations to Kennedy," *Aviation Week & Space Technology* 102 (April 21, 1975): 34.

20. Schechter, "The Concorde and Port Noise Complaints," 148–149; "Gov. Carey Voices Opposition to Concorde," *Aviation Week & Space Technology* 104 (March 1, 1976): 29.
21. Warren C. Wetmore, "State Noise Bill Would Block Concorde," *Aviation Week & Space Technology* 104 (March 1, 1976): 29–30.
22. Schechter, "The Concorde and Port Noise Complaints," 150–154. For more on the aviation industry's response to the PNYNJ's efforts to ban the Concorde, see Warren C. Wetmore, "Concorde Suit Raises Broad Issue," *Aviation Week & Space Technology* 104, (March 22, 1976): 27–29; "Concorde Suit In New York Called Pivotal," *Aviation Week & Space Technology* 104 (May 24, 1976): 29; "Concorde Ban Termed 'Discriminatory'," *Aviation Week & Space Technology* 106 (June 13, 1977): 38; "New York Ban On Concorde Continued," *Aviation Week & Space Technology* 107 (July 11, 1977): 24.
23. Erie, *Globalizing L.A*, 95, 99, 102; Paul D. Friedman, "Birth of an Airport: From Mines Field to Los Angeles International, L.A. Celebrates the 50th Anniversary of its Airport," *Journal of American Aviation Historical Society* (Winter 1978): 293; George S. Hunter," Los Angeles Moves to Meet Traffic Gain," *Aviation Week & Space Technology* 87 (July 17, 1967): 55.
24. "Legislative Action on Airport Noise May Follow West Coast Hearings," *Aviation Week & Space Technology* 72 (May 2, 1960): 51; "Real Estate Decline Pegged on Jet Noise," *Aviation Week & Space Technology* 74 (June 12, 1961): 41.
25. Erie, 102–103; Paul David Friedman, "Fear of Flying: Airport Noise, Airport Neighbors," *The Public Historian* 1 (Summer, 1979): 63–64.
26. Werlich and Krinsky, "The Aviation Noise Abatement Controversy," 83–85.
27. Ibid., 86–90.
28. Jeffrey M. Lenorovitz, "Airport Noise Controls at Issue in Trial," *Aviation Week & Space Technology* 112 (January 21, 1980): 28; "FAA Threatens Action Against Airport," *Aviation Week & Space Technology* 112(April 14 1980): 291; Jeffrey M. Lenorovitz, "FAA Irked by Orange County Action," *Aviation Week & Space Technology* 112 (May 26, 1980): 41, 43.
29. Werlich and Krinsky, 91.
30. For an overview of those arguments see "Shifting Aircraft Noise Liability to the Federal Government." *Virginia Law Review* 61 (Oct. 1975): 1299–1337.

31. Hardaway, 122–123; Werlich and Krinsky, 90–95.
32. Erie, 182–185.
33. Douglas Karsner, "Aviation and Airports: The Impact on the Economic and Geographic Structure of American Cities, 1940s–1980s," *Journal of Urban History* 23 (May 1997): 408–412.
34. Ibid., 418–20.
35. Ibid., 412–16.
36. Ibid., 420–26.
37. Ibid., 426–27.
38. The Minneapolis-St. Paul Metropolitan Airports Commission, "A Brief Report on Aviation Progress and Airport Problems in the Twin Cities," (January 30, 1957): 10–16, Box 5, Folder 1: Minneapolis-St. Paul Metropolitan Airports Commission, Subject files. Minnesota Historical Society, St. Paul, MN [hereafter MHS]; "Editorial: Big Aviation News," *The Minneapolis Star*, Friday, April 1, 1955.
39. Donald V. Harper, "The Minneapolis-St. Paul Metropolitan Airports Commission," *Minnesota Law Review* 55 (1971), 386–387, 406.
40. Atch to Ltr James T. Hetland, Jr., Chairman, Metropolitan Council to Representative Joseph P. Graw, Chairman, Subcommittee on Air Transportation, Metropolitan and Urban Affairs Committee, State of Minnesota, July 17, 1970, Box 5, Folder 1: Metropolitan Airports Commission, Subject Files, (MHS); Statement to the House Committee on Metropolitan and Urban Affairs by H. G. Kuitu, Executive Director, Minneapolis-St. Paul Metropolitan Airports Commission, February 1, 1973, Box 6, Minneapolis-St. Paul Metropolitan Airports Commission, Wold-Chamberlain Field Files, (MHS); Ltr J. H. Shaffer, Administrator, FAA, to Mr. Lawrence M. Hall, Chairman, Minneapolis-St. Paul Metropolitan Airports Commission, 18 January 1972, RG 237 Box 391, File: Minneapolis/St Paul Airport (Ham Lake Site); Betty Wilson, "New airport sites found," *The Minneapolis Star*, Thursday, Feb. 10, 1972; Ltr. J. H. Shaffer, Administrator, FAA, to Honorable Albert H. Quie, U. S. House of Representatives, 2 March 1972, RG 237, Box 391, File: Minneapolis/St Paul Airport (Ham Lake Site).
41. Donald V. Harper, "The Dilemma of Aircraft Noise at Major Airports," *Transportation Journal* 10 (Spring 1971): 22.
42. "On-going noise control efforts part of good neighbor policy," Aviation Advertising Supplement, *Minneapolis Tribune*, January 17, 1982.

43. The passage of the Airport Safety and Noise Abatement Act of 1979 also resulted in the development and adoption of new federal aviation regulations—FAR Part 150, Airport Noise and Land Use Compatibility Planning. Under FAR Part 150, the FAA provided grants to local airport owners and operators to develop noise abatement and land use plans.
44. Scott Skramstad, "History of Airport Noise Abatement," (Metropolitan Airports Commission, June 2002): 1–4; "Metropolitan Airports Commission," (Program Evaluation Division, Office of the Legislative Auditor, State of Minnesota, January 2003), 81–83.
45. See Metropolitan Council, *Aviation Policy Plan* (Metropolitan Council, adopted December 19, 1996), 25–31.
46. Ltr Walter T. Malloy, Director of Public Works, to Mrs. Fred Schrier, Maryland, Heights, Missouri, October 1, 1963, Raymond Tucker Papers, Series 3, Box 36, File: Lambert-St. Louis Municipal Airport, (University Archives).
47. Minutes, St. Louis Airport Commission, Thursday, September 24th, 1970; Minutes, St. Louis Airport Commission, Thursday, November 19, 1970, Cervantes Papers, Series Two, Box 11, File: Airport Commission, (University Archives).
48. See Bednarek, "Layer Upon Layer."
49. St. Louis Airport Environs Plan, Press Release: Environs Plan Completes First Year, January 9, 1981, Robert Young Papers, Box 42, File 900, (WHMC).
50. Prepared by Brown and Associates Public Relations for the Missouri-St. Louis Metropolitan Airport Authority, St. Louis Airport Authority, and Citizens for Lambert Committee," Lambert: For the Future" (1983), 13, Robert Young Papers, Box 42, File 901, (WHMC).
51. Ltr Harold Gaskin, President, Kinloch Merchant's Association to Congressman Robert Young, June 20, 1984, Robert Young Papers, Box 51, File 1036, (WHMC).
52. Janice Borgschulte, "Airport Group Oks Buyout of Business," *County Star-Journal*, April 3, 1985, Clipping in Robert Young Papers, Box 42, File 903, (WHMC); Ltr Charlton D. Clay, Mayor, Kinloch to Hon. Vincent C. Schoemehl, Mayor, City of St. Louis, November 16, 1990, Schoemehl Papers, Third Term, Box 8, File: Airport 1990, WHMC).
53. Ltr Charlton D. Clay, Mayor, Kinloch to Mr. Milt Svetanics, Mayor's Office, City of St. Louis, February 14, 1991, Schoemehl Papers, Third Term, Box 8, Airport 1991 (WHMC).

54. Ltr E. W. (Bill) Abram, Mayor, City of Bridgeton to Congressman Robert A. Young, U.S. House of Representatives, December 30, 1985, Robert Young Papers, Box 35, Folder 779, (WHMC).
55. "Culter: 'No Runways in Bridgeton'," *Bridgeton Air Defense News* 2 (January 1990); 1; Ltr Conrad W. Bowers, Mayor, City of Bridgeton to Bridgeton Residents, January 16, 1990, Schoemehl, Third Term, Box 9, File: Airport Misc.-Jan. 1990, (WHMC); Typewritten mss. "Statement from Vincent Schoemehl, Mayor of the City of St. Louis (regarding the Lambert Airport Master Plan)" Delivered by: Marie Boykin, Spokesperson for the Mayor, September 25, 1991, Henry the VIII Hotel (North Lindbergh), Schoelmehl Papers, Third Term, Box 7, File: Airport 1991, (WHMC).
56. Ltr Vincent Schoemehl, Jr., Mayor, City of St. Louis to Congresswoman Joan Kelly Horn, Washington, D.C., October 1, 1992, Schoemehl, Third Term, Box 19, File: Airport 1992, (WHMC).
57. Passenger Statistics, Lambert-St. Louis International Airport http://www.lambert-stlouis.com/flystl/media-newsroom/stats/ (accessed July 21, 2011).
58. William C. Barnard, "Airport Expansion Gets OK," *The Plain Dealer*, Thursday, September 26, 1968. The land acquisition battle between Cleveland and Brook Park began in late 1967 as Brook Park began a fight to purchase 54 acres of land adjacent to the airport that the city needed for runway extensions. After both cities bid on the land, with Cleveland winning the bidding war, Brook Park took its case all the way to the Ohio Supreme Court, where the city lost. "Suburb OK's Buying Land at Airport," *The Plain Dealer*, December 10, 1967; John Nussbaum, "Brook Park Action Challenges City's Title to Acreage," *The Plain Dealer*, January 10, 1968; "City Wins Contested Airport Land," *The Plain Dealer*, October 7, 1971.
59. Andrew M. Juniewicz, "Hopkins in Need, Jetport or No," *The Plain Dealer*, February 28, 1973.
60. For a sense of the political and finance issues associated with airport expansion in the 1970s, see Andrew M. Juniewicz, "150 airport home sales frustrated after 2 years," *The Plain Dealer*, February 25, 1974; _____., "'New' airport aiming for efficiency, ease," *The Plain Dealer*, March 24, 1974; _____., "Airport expansion blocked pending purchase of homes," *The Plain Dealer*, April 12, 1974; Charles Tracy, "Airport has anti-noise plan," *Cleveland Press*, October 19, 1976; Andrew Juniewicz, "Council bottleneck stalls Hopkins expansion,"

The Plain Dealer, March 6, 1976; "Council votes $26 million to speed up airport work," *Cleveland Press*, March 30, 1976; Andrew Juniewicz, "Funding is delayed on airport project," *The Plain Dealer*, July 20, 1976; John Nussbaum, "Suburbs, Hopkins near accord on plan for runway," *The Plain Dealer*, September 26, 1977; Charles Tracy, "City seeks new runway for Hopkins by 1980," *Cleveland Press*, January 26, 1978.

61. For a sense of the battles over noise and runway extensions in the 1980s, see Steve Lettner, "City tries to quiet Brook Park noise fears," *The Plain Dealer*, July 13, 1983; David Beard, "Hopkins runway show held to quiet noise fears," *The Plain Dealer*, September 16, 1983; _____., "7 western suburbs fight Hopkins plan," *The Plain Dealer*, December 8, 1983; V. David Sartin, "$8 million plan would add to ends of runway," *The Plain Dealer*, December 8, 1983; Jane M. Littleton, "Mayors to tell FAA runway plan is bad" *The Plain Dealer*, September 29, 1984; John F. Hagan, "Airport noise suit against city dropped," *The Plain Dealer*, March 27, 1985; Harry Stainer, "Hopkins offers anti-noise plan," *The Plain Dealer*, June 29, 1985; Eric Stringfellow, "More homeowners near Hopkins want buy-outs," *The Plain Dealer*, May 22, 1986; John S. Long, "Family will not relocate," *The Plain Dealer*, February 3, 1989; Sandra Livingston, "Hopkins' capacity crunch," *The Plain Dealer*, November 26, 1989.
62. Lou Mio, "Hopkins expansion plan axes 400 homes, I-X Center," *The Plain Dealer*, June 26, 1992; Benjamin Marrison, "NASA unit in path of Hopkins expansion," *The Plain Dealer*, June 27, 1992.
63. For more detail on the final agreement, see copy of "Settlement Agreement Between the Cities of Brook Park, Ohio, and Cleveland, Ohio," Vertical Files, Cleveland Public Library, Public Administration; News Release, Office of Mayor Michael R. White, Joint Press Release City of Cleveland – City of Brook Park, "I-X Settlement Agreement Finalized; Mayor White and Mayor Coyne Unveil Funding Plan" Vertical Files, Cleveland Public Library, Public Administration.
64. "Riveredge Township: The Encyclopedia of Cleveland History," http://ech.case.edu/cgi/article.pl?id=RT (accessed September 15, 2014); David Beard, "City explains Riveredge relocation plan," *Cleveland Plain Dealer*, November 17, 1983.
65. Beard, "7 western suburbs fight Hopkins plan."

66. V. David Sartin, "$8 Million plan would add to ends of runway," *Cleveland Plain Dealer*, December 8, 1983.
67. David Beard, "Port noise complaint: Even possibility of runway extension is fought," *Cleveland Plain Dealer*, March 10, 1985; David Sartin, "Aerospace center idea would hem in airport," *Cleveland Plain Dealer*, May 22, 1985.
68. "Riveredge Township: The Encyclopedia of Cleveland History," http://ech.case.edu/cgi/article.pl?id=RT (accessed September 15, 2014).
69. Al Thompson, "Airport project official works to help 'trapped' residents," *Cleveland Press*, October 26, 1972; Andrew M. Juniewicz, "150 airport homes sales frustrated after 2 years," *Cleveland Plain Dealer*, February 25, 1974.
70. Charles Tracy, "Hopkins starts anti-noise plans," *Cleveland Press*, January 2, 1981; David Beard, "Residents plan to fight Hopkins airport noise," *Cleveland Plain Dealer*, September 14, 1983; _____. "Hopkins runway show held to quiet noise fears," *Cleveland Plain Dealer*, September 16, 1983.
71. Beard, "7 Western suburbs fight Hopkins plan."
72. Quoted in Eric Stringfellow, "Sky is falling, or so it seems: Noise relief for residents near Hopkins still up in the air," *Cleveland Plain Dealer*, May 18, 1986; _____., "More homeowners near Hopkins want buyouts," *Cleveland Plain Dealer*, May 22, 1986.
73. Eric Stringfellow, "Sky is falling, or so it seems: Noise relief for residents near Hopkins still up in the air," *Cleveland Plain Dealer*, May 18, 1986.
74. John S. Long, "Family will not relocate: Says life with noise beats life apart," *Cleveland Plain Dealer*, February 3, 1989.
75. Frenchie Robles, "Neighborhood taking flight: Airport noise leaves deafening silence," *Cleveland Plain Dealer*, January 31, 1993.
76. Benjamin Marrison, "Airport's growing pains neighbors: 440 more houses to be demolished," *Cleveland Plain Dealer*, April 1, 1993.
77. V. David Sartin, "Making a return flight: Reunion brings folks back to airport neighborhood," *Cleveland Plain Dealer*, July 16, 1995.
78. Alison Grant, "An end for homes: Neighborhood near Hopkins Airport accepts news of speedier buyout with sadness, relief," *Cleveland Plain Dealer*, February 4, 1998; Brian E. Albrecht, "Neighborhood won't die easily," *Cleveland Plain Dealer*, October 4, 1998.

79. "Razing the I-X Center, Dick Jacobs' will and Kucinich's Integrity Now: Whatever happened to?..." http://blog.cleveland.com/metro/2010/04/razing_the_i-x_center_14_milli.html (accessed July 27, 2011).
80. Jim Calhoun, "Residents get second say on new runway," *Cincinnati Enquirer*, October 21, 1986; Tony Pugh, "Delhi residents question runway's fairness," *Cincinnati Enquirer*, October 23, 1986; "Delhi Township wins second runway hearing," *Cincinnati Enquirer*, November 16, 1986; Bob Musselman, "Runway's foes turn out in force for Ohio hearing," *Cincinnati Post*, November 26, 1986.
81. Dick Rawe, "Airport seeking federal funds for new runway," *Cincinnati Post*, June 1, 1988; Steve Kemme, "New runway could bring billions here," *Cincinnati Enquirer*, July 27, 1988; Dick Rawe, "Economy poised for takeoff," *Cincinnati Post*, July 27, 1988; Bill Robinson, "Delhi threatens court showdown over runway," *Cincinnati Post*, July 28, 1988; Crystal Harden, "FAA approves funds for new runway," *Cincinnati Post*, August 4, 1988; Sharon Moloney, "Delhi officials criticize airport runway study," *Cincinnati Post*, August 6, 1988; Ginny Hunter, "Runway fighters gain ally," *Cincinnati Post*, September 23, 1988; "Delhi sues over runway," *Cincinnati Post*, September 26, 1988; Ginny Hunter, "Runway on hold in court," *Cincinnati Post*, September 27, 1988; "Judge rejects Delhi's runway suit," *Cincinnati Post*, October 3, 1988; Ben J. Kaufman, "Judge OKs beginning new runway," *Cincinnati Enquirer*, October 4, 1988; Al Andry and Monica Dias, "Runway construction at airport to begin," *Cincinnati Post*, October 5, 1988.
82. Monica Dias, "Airport board OKs runway pact," *Cincinnati Post*, December 20, 1988.
83. Ken Wilson, "Township candidates seek airport board investigation," *Cincinnati Post*, September 5, 1991; Ben Kaufman, "First airport workshop held," *Cincinnati Post*, September 1, 1992.
84. "Luken eyes airport board changes," *Cincinnati Post*, September 16, 1992; Ken Wilson, "Luken: Put Ohio on airport board," *Cincinnati Post*, September 17, 1992; "Luebbers wants airport voice," *Cincinnati Post*, September 25, 1992; "County plans suit on airport," *Cincinnati Post*, October 3, 1992; Dan Horn, "County joins airport committee," *Cincinnati Post*, October 24, 1992; Ben L. Kaufman and Geoff Hobson, "Trio's letter complains about board," *Cincinnati Enquirer*, February 10, 1993; Jennifer Maddox, "Mann wants Voinovich to

assist in airport fight," *Cincinnati Post*, April 2, 1993; Monica Dias, "Airport meetings poorly attended," *Cincinnati Post*, April 17, 1993; Jennifer Maddox, "Airport spat goes to D.C.," May 6, 1993; "Seat on airport board pushed," *Cincinnati Post*, June 17, 1993.Kristi Bowden, "Airport board smooth voting," *Cincinnati Enquirer*, December 26, 1995; Monica Dias, "Ohioans will sit on airport board," *Cincinnati Post*, December 18, 1998.

85. Nancy Firor, "Battling airport noise: Abatement plans hinge on funding," *Cincinnati Enquirer*, Mary 21, 1990; Frederick Bermudez, "Routes altered to cut plan noise," *Cincinnati Enquirer*, October 31, 1990; Mike Turmell, "Jets drown out sounds of Oakbrook: Airport noise means closed windows, shaky floors on Parfour Court," *Cincinnati Enquirer*, May 6, 1991.
86. Patrick Crowley, "Airport expansion plan: Buy, destroy 214 houses," *Cincinnati Enquirer*, March 4, 1999; Monica Dias, "Runway land bill increasing," *Cincinnati Post*, September 21, 1999.
87. "Greater Cincinnati Northern Kentucky Airports – History," http://www.cvgairport.com/about/history2.html' (accessed March 1, 2012); "Why CVG Lost Half All Flights," http://news.cincinnati.com/article/20100524/EDIT03/5230393/Why-CVG-lost-half-all-flights (accessed March 1, 2012).

PART IV

Security: Hijackings, Hare Krishna, and September 11

Chapter Seven: Airport Security: Hijackers, Terrorists, Religious Groups, and the Constitution

When most Americans in the twenty-first century think about the issue of airport security, likely they do so within the context of the extensive measures adopted in the wake of the September 11, 2001, terrorist attacks. However, as shown in chapter one, airport security as a serious national security issue is one that reaches back at least to the late 1950s and early 1960s as airplane hijacking became an all too familiar part of the aviation scene. While local and federal officials reacted to each wave of hijackings from the 1950s onward, a series of hijackings between 1968 and 1973 proved particularly significant as they prompted the creation of the very basic airport security measures that remained familiar to air travelers until 2001. Though many at the time sounded an alarm as to the implications of such screenings for fourth amendment rights, for the most part, Americans seemed to accept the added measures as a needed price for aviation safety and courts generally upheld the security measures as constitutional.

As waves of hijackings waxed and waned over the years, so too did the attention paid to aviation security. Periodically, from the early 1970s through 2001, a particularly deadly or spectacular terrorist incident would prompt another period of examination of procedures and technologies. These actions also revealed the basic tensions behind efforts to provide for aviation security. First, there came frequent debates about who was responsible for aviation security—the federal government or the airlines and local airport officials. Second, discussions of airport security also

J.R. Bednarek, *Airports, Cities, and the Jet Age*,
DOI 10.1007/978-3-319-31195-1_8

revolved around how to balance a need for some level of security versus the convenience and comfort of passengers. Finally, how could officials balance security with fourth amendment rights to freedom from unreasonable search and seizure.

Airports became battlegrounds over not just fourth amendment rights, but first amendment activities as well. Beginning in the 1970s, a number of organizations, including religious groups, began to appear at major airports across the country distributing literature and soliciting donations. From the late 1970s through the early twenty-first century, airport officials' attempts to regulate the activities of such groups prompted court battles. Though lower courts often ruled in favor of the organizations, beginning in the early 1990s, higher court decisions began to uphold the regulations. And post-September 11, in a case involving the Los Angeles airport and the Hare Krishna, the California Supreme Court cited security issues in upholding a ban on the immediate solicitation of donations on airport property. Moreover, this and similar cases also explored the very definition of an airport, particularly was it a public space or, if public, what kind of public space it was.

Federal Versus Local Responsibility for Aviation Security: All Hijackings Are Local

Though hijacking was an international crime, it was also a domestic issue and many felt that the best way to tackle it was to stop it at the source—to prevent hijackers from getting on airplanes in the first place. While, as noted in chapter one, the Nixon Administration managed to prevail in terms of federal versus local responsibility for passenger screenings and airport security, the issue arose again every time events resulted in the proposed or actual implementation of enhanced security measures. In the mid-1980s, for example, following a number of high profile incidents, the FAA called for increased security measures at US airports. Enhanced security again became a sudden priority following the bombing of Pan Am 103 over Lockerbie Scotland in 1988 and after TWA 800 exploded shortly after take-off in late 1996,[1] though the latter proved not to be a terrorist attack. In those latter two cases, attention focused on technology capable of detecting even a small amount of explosives. Though explosives detection technology was becoming more available, it was still expensive and difficult to operate. Airlines and local airport managers pushed back against federal mandates to install such equipment as well as against requirements to implement other potentially costly security measures, such as 100 % passenger-baggage matches on all flights, domestic as well as international. In protesting, the airlines cited not only the costs to themselves, which might put them at a competitive disadvantage to foreign air carriers operating in the USA, but also the potential cost in passenger inconvenience. Eventually, Congress shifted a small measure of the responsibility and cost for security to the FAA, when it authorized the FAA to purchase and install the new detection equipment. For the most part, though, security remained an airline and local responsibility until after September 11, 2001.

The mid-1980s witnessed a heightening of international tensions and a number of very violent attacks against aviation targets worldwide. In contrast to the 1970s, all the incidents happened outside the USA. They included several hijackings as well as a mass shooting at the airport in Rome, Italy. In addition, the FAA ordered stronger security measures at US airports, particularly those handling international flights, after the USA launched air strikes against Libya following a bombing, linked to the government of Muammar Qaddafi, at a West Berlin nightclub. Most of the attention, though, focused on security on international flights and at airports outside the USA. The Reagan Administration reinstated the air marshal program and the FAA placed armed marshals on certain international flights. The State Department focused on negotiating additional bilateral agreements that would allow armed marshals on more international flights as well as increased joint anti-terrorist measures at non-US airports. Additionally, the FAA sent security experts to a number of overseas airports to help US carriers improve security at their facilities within those airports.[2]

Despite the focus on overseas airports and on major international airports in the USA, following the recommendation of a departmental safety review task force, in 1986, the Department of Transportation (DOT) directed the FAA to take measures to tighten security at all US commercial airports. The recommendations were as follows: end the exemption for airline and airport personnel from passing through the metal detector before entering secure areas of the airport and require them to submit any bags to the X-ray machine; limit the number of doors or other openings between the secure and non-secure areas within the airport; and separate unscreened passengers from commuter or private aircraft from screened airline passengers.[3]

Airport managers responded positively to these recommendations, but airline executives objected. Under the user-pay system, most of the additional costs for airport security were passed on from the airports to the airlines. Before deregulation, the airlines could petition the Civil Aeronautics Board for fare increases to cover the costs. In the immediate post-deregulation period of downward price pressures, it was harder to raise fares to cover security costs. Further, the concerns over security caused many Americans to cancel or postpone international travel.[4] While airport executives meeting at a conference in 1986 listed security as among their top concerns, a group representing the nation's airlines suggested that the focus on threats was out of proportion to reality. Richard Lally, the assistant vice president for security for the Air Transport Association, bluntly suggested that the heightened media concern over the safety of flight was overblown. He asserted that the most dangerous part of an airline trip was not that part spent by a passenger on an airliner, but the part they spent in their vehicle traveling to the airport in the first place. He noted that the number of hijackings in 1985 (four) was the lowest number since 1976. He also criticized the media for concerns over a so-called "plastic gun" that could make it past airport metal detectors. He pointed out that the only such weapons then in existence had more metal in them than "the Saturday Night Specials so prevalent in the United States." He did, however, endorse continued FAA research on an explosives detection devise. He called that the "one tool that is needed."[5]

In 1986, explosives detection technology was still very much in the research and development stage. Both Westinghouse and Science Application International Corporation (SAIC) worked under an FAA grant program to develop devices based on Thermal Neutron Activation (TNA) Explosive Detection systems. Though the Westinghouse and SAIC systems differed, TNA technology essentially bombarded luggage with neutrons that reacted with the nitrogen used in explosives causing the nitrogen briefly to emit a specific type of radiation which the devise then detected. A third company was working on a chemiluminescence system—a gas that would react in the presence of nitrogen. Though development continued throughout the 1980s, all explosives detection technology remained largely experimental into the early 1990s. In the aftermath of the bombing of Pan Am 103 in December 1988, however, Congress and the FAA sought to accelerate the testing and deployment of TNA technology and, once again, tighten airport security.[6]

While the FAA ordered increased security on US planes flying between the USA and Europe, it did not require heighten security measures for foreign aircraft flying the same routes. Such action necessitated negotiations between the USA and the foreign airlines' countries of origin. The foreign carriers and the governments of their countries of origin (many of which had an ownership stake in the airline) resisted FAA oversight of their security measures as an infringement on their sovereign rights. The issue of security on foreign carriers was significant for two reasons. First, since deregulation, the percentage of US passengers flying overseas on foreign carriers had risen from close to 35 % to almost 50 % as the US government, in an effort to promote competition on overseas routes, had negotiated agreements allowing foreign carriers more access to the US market. Second, US carriers argued that the costs of the new security measures put them at a competitive disadvantage with the foreign carriers.[7]

Once it became clear that a bomb had brought down Pan Am 103, attention focused again on acquiring technology capable of detecting explosives in checked and carry-on baggage. In early 1989, the FAA had on order six test TNA devices at a cost of $1 million per machine. It proposed experimenting with the new machines at a number of US airports deemed at highest risk due to the number of international flights. After the British concluded that the Pan Am 103 bomb had been concealed within a tape recorder, the FAA ordered airlines to inspect all electronic devices brought on board international flights in both carry-on and checked bags. Overall, though, while security measures were generally greater, there was little uniformity. US carriers had the strictest security regulations, while the security practices of foreign carriers varied. Further, many of the regulations covering US carriers, including those involving electronic devices, only applied to flights from Europe and the Middle East to the USA, not on flights from the USA to overseas destinations. And by the end of the year concerns over radiation had led to delay in the testing of the TNA machine at Miami International (and the machine was eventually removed) and at London's Gatwick. In late 1989, only the machine at Kennedy was in operation. Eventually, another machine was tested at government-owned Dulles International Airport.[8]

In November 1990, following a report by the President's Commission on Aviation Security and Terrorism, and almost two years after the Pan Am bombing, Congress passed the Aviation Security Improvement Act of 1990. It called for a number of changes, including background checks on all newly hired air carrier personnel (both direct hires and contract employees, especially those responsible for screening passengers and baggage) and the creation of a process to notify the public of credible threats to aviation security. It also, however, placed a moratorium on the purchasing of additional TNA machines. At the time of the bill's passage, FAA testing of the technology had shown that the machines registered a higher false alarm rate (15–20 %) than predicted, especially when set to detect very small amounts of explosives. (The Pan Am 103 bomb, for example, contained only one pound of explosives.) The machines also operated more slowly than had been promised. US airlines strongly opposed mandating such equipment, which spokespersons for the airlines described as expensive, bulky, and unreliable. Both independent and airline-sponsored investigations concluded that the technology needed further refinement and called for additional research and development. A presidential commission and Congress agreed and the new legislation called both for more research and development and prohibited the FAA from purchasing additional equipment until testing certified its reliability.[9]

After the passage of the Aviation Security Improvement Act, airlines not only objected to the use of the TNA technology but also complained about the high costs of many of the other recommended actions, including pre-employment screening of personnel and systems to safeguard access to secure areas of the airports. In response, the FAA scaled back those requirements. And problems persisted with explosives detection equipment. Though the FAA had certified a new device, the CTX 5000, it was still being tested at US airports where the airlines using it reported slow operating speeds and high rates of false alarms. These machines, like those of the first generation, also cost $1 million apiece. The FAA and the airlines also came under criticism for failure to develop a program to check cargo before it was loaded, especially on aircraft also carrying passengers.[10]

The bombing of the World Trade Center in New York City in 1993 followed by the destruction of the federal building in Oklahoma City in 1995 once again focused the attention of the American public on the problem of terrorism—both foreign and domestic. That attention turned to aviation during the summer of 1996. On July 17, 1996, TWA 800, a Boeing 747 enroute to Europe, exploded shortly after take-off from New York's Kennedy Airport. Although eventually the National Transportation Safety Board determined that a spark from short circuit in the wiring in one of the plane's fuel tanks had caused the explosion, initially the government could not rule out an act of terrorism.[11] Immediately, both the FAA and the airlines came under intense criticism for what many saw as a fatally weak aviation security system.

After the destruction of TWA 800, President William Clinton appointed a commission to investigate and ordered it to report back quickly.

The commission, chaired by Vice President Al Gore, issued a preliminary report with a list of recommendations in early September 1996. At the same time, Congress was at work on an anti-terrorism bill. That legislation had actually begun more in response to the Centennial Olympic Park bombing on July 27, 1996, than TWA 800. But with both those events in the summer of 1996, Congress and the Clinton Administration were anxious to take quick action in an important election year. However, rather than folding the recommendations on aviation safety into the proposed anti-terrorism bill, the recommendations became part of the Federal Aviation Administration Reauthorization Act. Among the recommendations adopted in the legislation was a requirement for airlines to match passengers and baggage on all flights, not just international flights. Also, the bill authorized the creation of a computerized profiling system to determine which passengers should be subject to greater scrutiny. And it required criminal background checks on all airline employees, including the contract employees responsible for security. Despite strong opposition from the airline industry, which worried about high costs and passenger inconvenience, Congress passed the bill into law and President Clinton signed it on October 9, 1996.[12] The Federal Aviation Administration Reauthorization Act was not exclusively about security, but it was important for the way it began the reshaping of the relationship between the federal government, the airlines and local airport authorities in relation to security.

Prior to the 1996 legislation, local airport officials and the airlines had been responsible for financing all security measures at the nation's airports. For example, either the airlines hired the personnel and purchased the equipment directly or they contracted with local airport authorities to handle those tasks. Either way, airlines (and ultimately American airline passengers in the forms of higher fares or fees) paid the costs associated with security. Now the cost shifted, at least in part, to the federal government. The 1996 legislation made the FAA responsible for certifying, purchasing, and deploying the new, expensive bomb detection equipment. It authorized the purchase of 54 certified screening devises, an expansion of the FAA's security forces, and FAA training of personnel to operate the new baggage screening machines. Furthermore, the legislation called for a study of how increased security measures might be financed in the future.[13]

The new legislation, however, did not provide federal funding for the proposed universal passenger-baggage match system. Airlines argued that they did not have the personnel or equipment to implement such a system. Further, such a system, if required, could lead to long delays and canceled flights. Airlines also continued to complain about the costs of implementing the background checks which they called excessively costly and time consuming. Such lengthy procedures, they claimed, would make it difficult for airlines to hire well-qualified employees.[14]

By fall 1998, the FAA reported great success in the deployment of the explosives detection equipment.[15] Despite that upbeat report, within a year, it became clear that the airlines were not fully using the new equipment. An audit of machine use found that while the machines were designed to handle up to 225 bags per hour, most did not handle that many bags per day. The airlines again cited high rates of false alarms, this time because the technology frequently mistook food, often packed by passengers, for explosives. The FAA admitted that it had not tested for food-related false alarms due to issues of spoilage over the several-days-long test period. But the FAA claimed that a computerized passenger screening system then under development would create a data base to determine which passengers should have their luggage subjected to the most intense form of inspection. To simply require the examination of all bags, the FAA conceded, would entail lengthy delays and passenger inconvenience due to longer wait times as well as to possible missed flights.[16]

Airport Security, the Fourth Amendment, and the Courts: Paving the Road to the "New Normal"

There were those who challenged the security system devised in the wake of the hijackings of the late 1960s and early 1970s. Officials at the airlines and airports, as noted, balked at the costs, both in terms of machinery and personnel as well as passenger inconvenience. Moreover, others doubted the constitutionality of the system, citing the fourth amendment protection against unreasonable search and seizure. The anti-hijacking security system, though, came about at a time when the courts were in many ways lowering the requirements necessary for warrantless searches. The standard had been probable cause. By the early 1970s, the courts were upholding searches based on reasonable suspicion of criminal activity. Courts also upheld searches in certain cases in which the government had a significant interest in protecting the general welfare. Therefore, though many expressed doubts about the use of the magnetometer and the search of carry-on baggage, US courts generally ruled in favor of these searches. The American public also seemingly accepted the new procedures as necessary to prevent or at least deter further hijackings. The fact that hijackings decreased following the adoption of the system helped. Though controversy still surrounded those cases where airport searches resulted not in the arrests of suspected hijackers but of individuals carrying illegal drugs, the American public for the most part accepted the idea of being searched as a prerequisite for boarding a commercial airline flight.

As noted in chapter one, the passenger screening process that became familiar to Americans in the last decades of the twentieth century began in 1969 as a voluntary program. Only a few airlines participated and they did so at only a few select airports deemed as high risk. In addition to that voluntary program, US customs agents inspected the carry-on bags of all passengers boarding international flights flown by US carriers. Coming at a time when hijackings were a major problem, many justified the measures as necessary to protect passengers and prevent air piracy. Others, however, questioned whether these measures violated the protections against unreasonable searches afforded under the Fourth Amendment to the constitution.[17]

The Fourth Amendment protects individuals from unreasonable searches conducted by government agents. In defense of the airport screenings on domestic flights, advocates argued that the airlines, which were private corporations and not agents of the government, conducted the searches. Therefore, the Fourth Amendment did not apply. However, critics charged that at that time the government still regulated the airlines and governments or government agencies (such as a public authority) owned most airports. And they pointed to case law dealing with the 14th amendment (equal rights) that supported the idea that private leasees on government-owned property should be held to the same standards as would apply to the state. However, the legal test in those cases—the “special relationship” between the leasee and the government—that applied in civil rights cases had not yet been applied to airports and airport searches.[18]

Those supporting airport searches further argued that the Fourth Amendment protected individuals only from *unreasonable* searches. There had long been exemptions to the fourth amendment requirement for a warrant and contemporary case law had expanded the exceptions that allowed for warrantless searches. Under certain circumstances, some argued that the same logic could also apply to airport searches. Traditionally, warrantless searches had been allowed under two conditions. First, authorities could conduct a search if the person involved voluntarily consented to the search. Such an exception, critics countered, did not apply to airport searches as a person could be denied boarding if they refused the search. Therefore, a certain measure of coercion was involved. Second, border agents had been specifically allowed by law to conduct searches of cargo ships if they believed the duty had not been paid on the materials being shipped. However, the law allowed for the searches only if the custom agents had reasonable cause to believe that the duty had not been paid on the ship's cargo and critics concluded searches of all passengers boarding international flights could not be justified under that precedent. In addition to those traditional examples, more recent exemptions to the warrant requirement—which allowed so-called administrative searches "conducted as part of general regulatory schemes in furtherance of legitimate administrative purposes"—did seem to apply to airport searches, though again some disagreed.[19]

Proponents of airport searches pointed to the fact that the voluntary screening system involved a multi-step process. First, the DOT had developed a so-called "hijacker profile." Airline employees observed passengers to identify those fitting the behavioral characteristics specified by the profile. Those singled out by the profile were then asked to pass through a magnetometer; if the magnetometer suggested the presence of a weapon, the person would be asked to submit to a search. Contemporary case law suggested that such a multi-step process met fourth amendment requirements.[20] However, many airlines steered all passengers through the magnetometer whether they fit the profile or not, though the magnetometer was not always operating. Nonetheless, such an indiscriminate use of the magnetometer would become an issue in court cases.

In an important decision in 1971, the District Court for the Eastern District of New York in *United States v. Lopez* ruled favorably on the validity of the hijacker profile, stating that it was accurate enough to establish probable cause. That ruling opened the door to the application of the Supreme Court decision in *Terry v. Ohio* to the airport searches. The *Terry* case involved a warrantless search conducted by a police officer. In *Terry*, the police officer had observed a number of individuals loitering outside a store. He approached the group. One of the individuals mumbled some words that led the officer to fear that a weapon would soon be used. In response, he conducted a frisk search and discovered the weapon. The Supreme Court denied the motion to exclude the weapon as evidence. The court argued that the policeman, based on his experience and training, had a reasonable suspicion that the defendant had a weapon and posed an immediate threat to the policeman and others. Therefore, a limited search—the frisk or pat-down of the outer clothing worn by the defendant—was a reasonable search under the Fourth Amendment. In the case of airport searches, it was argued, the profile was accurate enough to create a reasonable suspicion and allow for the use of the magnetometer.[21] Thus, the strict multi-stage process—profile, magnetometer, physical search—could be justified under existing case law. However, the indiscriminate search of all passenger carry-on baggage or the requirement that all passengers pass through the magnetometer could not. More evidence was needed to prove that the government's interest in preventing hijackings was compelling enough to allow for an exemption to the requirements under the Fourth Amendment.[22]

Following the decision in 1972 to make universal airport searches mandatory and anticipating the 1973 implementation deadline, a number of legal scholars again examined their constitutionality. Case law since 1971 had generally continued to support the constitutionality of airport searches. Further, courts specifically ruled on the use of the magnetometer. In two cases, *United States v. Epperson* and *United States v. Slocum*, courts argued that the government's interest in preventing hijackings justified the routine use of the magnetometer on all passengers. One critic, however, argued that the routine use of the magnetometer made it an indiscriminate dragnet and patently unconstitutional. He also found unconstitutional the mandatory screening of all carry-on baggage when that search was not triggered by the profile and a positive indication from the magnetometer. And another critic objected to the *Terry* and *Lopez* decisions on the grounds that the courts should carve out a specific exemption to the warrant requirement if continued airport searches were to be allowed.[23]

In June 1973, however, the US Court of Appeals, Ninth Circuit, issued a key ruling in *U.S. v. Davis* which answered the charge that *Terry* and *Lopez* did not justify airport searches. The Court of Appeals did find the analogies between those cases and airport searches defective, but ruled that the government had an overwhelming interest in protecting passengers from the dangers of hijacking. Searches aimed at uncovering weapons or explosives before someone could board an airplane constituted a reasonable search as long as the person involved had the right to avoid the search by choosing not to board the aircraft and leaving the area.[24]

The court decision opened the door for universal screening of airline passengers. Though some critics continued to question the validity of airport searches, they generally focused only on those cases where the searches turned up evidence not of a potential hijacking, but of a violation of another law, most frequently drug laws.[25] Perhaps most importantly, most Americans seemingly accepted the new security system and it became a familiar part of the air travel process. As practiced before September 11, 2001, it seldom resulted in much delay or inconvenience, except when periodically, though temporarily, more zealously applied immediately following successful terrorist attacks on airliners. Most of the criticism involved complaints that the system was too lax, not too extreme. That proved especially true after September 11, 2001.

September 11 and Aviation's New Normal

On the morning of September 11, 2001, 19 terrorists, working in teams of 4 and 5, hijacked four US airliners—American Flight 11 and United Flight 175 out of Boston Logan; American Flight 77 out of Washington Dulles; and United Flight 93 out of Newark. American 11 struck the North Tower of the World Trade Center in New York City at 8:46 a.m. local time. Seventeen minutes later, at 9:03 a.m., United 175 flew into the South Tower. Everyone aboard both planes and an unknown number in the buildings died instantly. At 9:37 a.m., American 77 hit the Pentagon in Washington, D.C. Again, all aboard the plane and many in the building died on impact. Shortly thereafter, the Cleveland Air Route Traffic Control Center heard a radio transmission from United 175 indicating there was a bomb on board and that the plane was returning to the airport in Newark. It is unlikely that the passengers heard that announcement. Instead, a number of passengers had received cell phone calls and, aware of the events that morning, decided to fight back again the hijackers. At 9:57 a.m., a group of passengers charged the cockpit. Voice recordings indicate that the hijackers then decided to down the aircraft immediately. United 175 crashed shortly after 10:00 a.m. near Shanksville, Pennsylvania.[26] September 11, 2001, like December 7, 1941, became a date etched in US history.

Within months of these terrorist attacks against the USA, the longstanding private and local responsibility for thwarting attacks against civil aviation came to an end. In the aftermath, many Americans expressed surprise even shock at the fact that private companies with contract employees handled airport security. Most screeners earned little more than minimum wage and many companies experienced a near 100 % yearly turnover in personnel. Both the airlines and the security companies had strong incentives to keep costs at a minimum. Further, many companies continued to hire individuals as screeners without completing the FAA required background check.[27]

Perceived weaknesses in existing airport security programs led to a short but heated debate in Congress. Critics had long charged that the private screeners were poorly trained and poorly motivated. A Government Accounting Office (GAO) audit in 2000 deemed the performance of most screeners as not satisfactory. The GAO also pointed out that screeners in other countries earned much better wages and worked as screeners for a longer period of time. There was agreement that the security system needed change and improvement. The debate centered on how best to accomplish those goals. Initially, Secretary of Transportation Norman Mineta doubted whether federalizing the screening workforce—an idea that dated back to the 1970s—would be approved. However, he also noted that the airlines were facing tremendous economic pressures and would find it difficult to finance a new system. In Congress, the debate focused on two rival proposals. The Senate voted overwhelmingly (100–0) to approve a bill federalizing the screening workforce. The House, however, divided along partisan lines. Democrats favored a bill which, similar to the Senate bill, would federalize the screener workforce. House Republicans proposed a bill that would still have had private companies employing the screeners, but would place the private security companies under contract to the DOT, following the example set by European countries.[28]

Under pressure to act before the Thanksgiving holiday travel season, a compromise was finally worked out calling for the year-long phase-in of a federalized screener workforce at the nation's airports. The screeners would work for the DOT's new Transportation Security Administration (TSA). However, the bill also provided for the potential continuation of private, contracted screening services. Up to five airports were allowed to essentially opt out of the new system and have private companies, working under federal supervision, handle passenger screening as part of a trial program. After two years, all airports could gain approval to have private companies handle security. The private screeners, however, would work under government contract and heightened federal oversight.[29] Although the proposed Aviation and Transportation Security Act allowed for the limited reprivatization of security, for the most part, airports have continued to use the direct services of the TSA. As of 2014, only 18 airports (out of the more than 450 airports falling under the responsibility of the

TSA) had opted to switch to private security firms working under TSA contract.[30] However, whether handled by government employees or government contractors, security at the nation's airports was no longer strictly a local and private responsibility.

The Aviation and Transportation Security Act passed on November 18, 2001. It created the TSA and gave it the job of hiring a security workforce for all the major commercial airports in the USA within one year and to provide the personnel and equipment to screen all baggage by December 31, 2002. In March 2003, the TSA moved from the DOT to the newly created Department of Homeland Security, as part of a massive expansion of the federal government in response to the threat from international terrorism.[31]

After the September 11, 2001, terrorist attacks, many began talking about a so-called "new normal," the feeling that Americans would have to accustom themselves to the notion of their perpetual vulnerability to international terrorism. Where the "new normal" became most obvious was at the nation's airports. Particularly in the immediate aftermath of the attacks, American airline passengers found themselves in airports patrolled by uniformed and armed military personnel. Identification was scrutinized. Pat downs were frequent. And although the most stringent of the security measures eventually faded, the days of arriving at the airport 20 minutes before your flight were gone. In the second decade of the twenty-first century, only ticketed passengers are allowed past security into the gate area. Both coats and shoes must be removed; all baggage is screened. And for the most part Americans, for the sake of greater security (real or perceived), have adjusted to the new normal at the airport.

Airline Security and the First Amendment: Redefining the Airport and Security

In the 1980 movie, *Airplane!*, the leading man, Robert Hays, is shown fighting through crowds at an airport trying to reach his girlfriend. Along the way he encounters a number of people blocking his way including a man in a saffron robe holding a flower. Hays immediately punches him out. Audiences cheered. That particular gag referenced what many Americans viewed as a problem in a number of public places by the late 1970s and early 1980s—solicitations by members of the Hare Krishna (International Society for Krishna Consciousness [ISKCON]) movement. Beginning in the 1970s, officials responsible for parks, public monuments, airports, and other places frequented by Hare Krishna members struggled to deal with a public interaction many Americans found at least annoying, many even threatening.

In response to complaints about members of ISKCON and others practicing what many saw as aggressive panhandling techniques, a number of airport officials across the country, including those in the cities of Los Angeles and New York, forbade the distribution of literature or the solicitation of money by any group at their respective major airports. While a number of organizations challenged the bans, ISKCON's challenges garnered the most attention. A case involving ISKCON and the PNYNJ, which began in the 1970s, went all the way to the US Supreme Court and proved very significant. The decision hinged largely on a reconsideration of the public forum doctrine and the definition of what constituted a public versus a non-public forum. By establishing a very narrow definition of what constituted a public forum, a majority on the court defined airports as non-public.

The court cases, thus, touched on the definition of an airport. Was it public or private? Was it strictly a transportation facility? Was it like a park or a street? Was it like a shopping mall? In most cases, the courts viewed airports as simply transportation facilities and not like parks or shopping malls, both of which had been ruled as some type public forums. By strictly constructing airports as transportation facilities, the Supreme Court, as well as some lower courts, more easily allowed for restrictions on free speech. At the center of the arguments was the public form doctrine.

The initial establishment of the public forum doctrine is usually credited to a Supreme Court case in 1939, *Hague v. CIO.* In that case, the court ruled against ordinances restricting free speech or free speech-related activities (i.e., leafleting, marches) in public spaces. In the years that followed, the Court then established what was known as the vagueness doctrine as a test of breach of peace or disorderly conduct ordinances or statutes seeking to control free speech activities in public areas. The court ruled unenforceable such laws that were overly broad in their wording or gave "standardless" discretion to local authorities to determine what constituted a breach of peace or disorderly conduct. Then in 1965, legal scholar and first amendment expert Harry Kalven wrote an article for the *Supreme Court Review* in which he argued that streets, parks, and other public places constituted a public forum where citizens had a right to exercise free speech. As such, any effort to regulate access had to pass a strict test of reasonableness.[32]

From the 1960s through the 1980s, however, the Supreme Court issued a series of decisions that gradually narrowed the definition of a public forum and upheld local efforts to restrict free speech in certain types of spaces. The most important case in this series of decisions was *Perry Educ Ass'n v. Perry Local Educators' Ass'n*. In that case, the Court first enunciated a categorical approach to the issue of what constituted a public forum. In the majority opinion, Justice Byron White argued that there were three types of forums: traditional public forums, designated public forums, and non-public forums. Traditional public forums were those places long established by tradition or government action as appropriate venues for the exercise of free speech rights. Such places included public streets and parks. A designated public forum was a publicly owned place in which the state had allowed free speech activities. However, just because permission had been granted once did not mean that the permission could not be rescinded. Regulation of free speech in both traditional and designated public forums had to be reasonable, narrow, and had to reflect a legitimate state purpose. The third category outlined in *Perry* was the non-public forum. The court ruled that the state could impose broader restrictions on free speech in such venues; however those restrictions had to be judged as reasonable. This definition of a non-public forum figured prominently in *United States v. Kokinda* in which the decision upheld a Post Office regulation prohibiting solicitation on the sidewalk in front of the Bowie, Maryland, post office. The court majority held the sidewalk as a non-public forum and the regulation as reasonable.[33]

After complaints from passengers, airport operators began in the 1970s to seek ways to curtail or ban outright organizations, including religious groups, from soliciting funds and distributing literature at their facilities. ISKCON aggressively challenged these bans, often arguing that the regulations infringed on the freedom of not only speech but also religion, as members of the Hare Krishna form of Buddhism were required to perform a public act of begging. In most cases, state courts overturned those bans.[34] In 1992, however, the US Supreme Court issued a final ruling in a long-standing case involving ISKCON and the PNYNJ that partially supported port authority actions.

The struggle between the PNYNJ and ISKCON began in the mid-1970s when the port authority issued regulations banning both the distribution of literature and the solicitation of funds in the region's airports. After ISKCON prevailed in lower courts, the litigation reached the federal district court of appeals, which overturned a federal district court ruling and decided in favor of the PNYNJ. The Supreme Court was asked to review that decision.[35] The Supreme Court issued a split decision that further complicated the issue of what constituted a public forum and to what degree free speech could be regulated in such a setting. While a majority of justices agreed that an airport was a non-public forum and that airport operators could restrict the solicitation of funds, they disagreed on whether or not a ban on the distribution of literature was allowable. Thus, the case resulted in multiple written rulings, the most important of which were the majority decision, written by Chief Justice William Rehnquist, upholding the ban on both solicitation and distribution, and two concurring decisions, the first by Justice Sandra Day O'Connor, upholding the ban on solicitation, but not on distribution of literature, and the second by Justice Anthony Kennedy, rejecting the definition of airports as non-public, but agreeing that the ban on solicitation was reasonable.

The Rehnquist decision began with a discussion of the public forum doctrine and the definitions of traditional, designated, and non-public forums. In declaring airports non-public forums, Rehnquist, for the court, argued that since airports were relatively new, only appearing in the twentieth century, they could not be considered traditional public forums. Second, given the extensive litigation involving efforts by airport operators (governmental entities) to restrict solicitation and the distribution of literature, it could be argued that governments had never intended to designate airports as public forums. Therefore, airports should be considered non-public forums and bans on solicitation and the distribution of literature need only pass the "reasonableness" test.[36]

Voting with the majority on the ban on solicitation of funds, but issuing a separate decision on the issue of the distribution of literature, Justice O'Connor agreed that airports were non-public forums, but argued that only the regulations against the solicitation of funds passed the reasonableness test. Bans on the distribution of literature did not. She argued that airports, having rented space to a number of commercial establishments such as shops and restaurants, were functioning more like shopping malls. Given that airports, thus, were not single-purpose transportation facilities, she saw nothing reasonable in the restriction of free speech rights inherent in the ban on the distribution of literature.[37] And in his own separate decision, Justice Anthony Kennedy rejected the definition of an airport as a non-public forum, but agreed that the ban on solicitation of funds, but not on the distribution of literature, passed the reasonableness test.[38] Because of the difference between the three decisions, the Supreme Court's ruling only upheld the ban on solicitation, not that on the distribution of literature.

A main issue in these cases revolved around whether or not airports were public forums and in making that determination the courts had to accept a certain definition of what an airport was. Clearly, in most decisions favoring restrictions on the activities of groups at airports, the courts viewed airports as strictly transportation facilities. In the Rehnquist decision, as noted, the chief justice further stated that the newness of airports precluded them from being considered traditional public forums. And given the litigation concerning attempts to restrict free speech, he argued that airports could not be considered designated public forums either.

Yet in rejecting the idea airports could be either traditional or designated public forums, the courts failed to examine carefully the history of airports in the USA. During the 1920s, the first full decade of airport construction in the USA, and even into the 1930s, several cities used their powers to purchase parkland in order to establish their airports. Many of these cities either did not have the power to establish airports or, if enabling legislation had been passed, found the park powers far easier to exercise than the power to establish airports. And in many ways, these earliest of airports did resemble parks, as they were basically large, flat grassy areas. Preparation often included the clearing of trees or the grading of landing areas, but otherwise many of the first airports were relatively simple to engineer. Providing for sufficient drainage and electrical service seemed to the most daunting technological challenges. In several cases examining the validity of cities using their powers to purchase park land in order to establish airports, the courts sided with the cities because of the similarity between parks and airports.[39]

Though the technology of airports became more complex during the 1930s, in many ways, they continued to function much like parks, in that people came to the airport for recreational activities. In the late 1920s and well into the 1930s, for example, many people went to the airport not as passengers on the early airlines, but simply to watch airplanes take-off and land. Airports also hosted air shows to draw crowds or built observation decks to encourage casual gawkers. Only after World War II did the idea of the airport as strictly a transportation facility firmly take hold.[40] While it was true that by the early 1990s airports hardly functioned as parks, an examination of their earlier history might challenge the notion that they were neither traditional nor designated public forums.

In determining whether or not airports constituted public forums, courts also examined the similarity between airports and shopping malls. When Justice O'Connor likened airports to shopping malls in the 1992 ISKON decision, the extensive "malling" of America's airports was a relatively new phenomenon, though certain retail activity had long been a feature at airports. However, over the next decade, airports all over the country followed the example of the Pittsburgh International Airport and its AirMall. The idea of an airport being like a shopping mall, thus, became more plausible over time. In fact, ISKCON later used the argument that airports were functioning like shopping malls in challenging a ban placed on the solicitation of donations and the selling of literature at the Miami International Airport in the late 1990s. In that case, though, the appeals court for that region upheld a ban on solicitation. ISKCON appealed to the US Supreme Court, but in 1999, it refused, without comment, to review that decision.[41]

The Supreme Court had ruled against early attempts to restrict first amendment activities in shopping malls, viewing them as the equivalent of such traditional public forums as city sidewalks, but the court became more conservative in the 1970s. In 1972, for example, the court allowed a shopping mall owner, based on property rights, to restrict the distribution of anti-war literature within the mall. In 1980, the court ruled against a mall owner seeking to ban petition gatherers, but reaffirmed its 1972 ruling that the First Amendment to the US Constitution did not guarantee access to a shopping mall. Rather, the court directed, states should determine whether or not their constitutions allowed for such access. In much the same way, access to airports became more of a state/local issue than a federal issue.[42]

The nature of state constitutional protections of certain rights played an important role in the outcome of the conflicts between ISKCON and airport officials in California. The history of conflict between religious groups and the Los Angeles airport commissioners began in the early 1980s. In 1983, the Board of Commissioners of the Los Angeles airport adopted a resolution aimed at banning all first amendment activities in the terminal area of the Los Angeles International Airport (LAX). Shortly thereafter a member of Jews for Jesus was stopped from distributing literature in the LAX terminal. That individual and Jews for Jesus brought suit against the commissioners. While the case brought up the question of whether or not the airport was a public forum, that issue was not addressed in the unanimous decision. Instead, in 1987, the US Supreme Court struck down the action of the commissioners as unconstitutional as it was overbroad.[43] The ruling, though, did not overturn other lower court decisions allowing certain limited restrictions on soliciting by religious groups.[44]

Following the 1992 US Supreme Court decision in the ISKON case, Los Angeles officials again attempted to restrict solicitation at the airport, and in 1997, the city passed a new ordinance banning all solicitation. ISKCON challenged the constitutionality of the ban and succeeded in getting an injunction against the enforcement of the ordinance. Following the terrorist attacks in September 2001, a temporary ordinance went into effect, mandating that those wishing to solicit funds at the airport apply for a permit and restricting such groups to certain areas of the airport. ISKCON again challenged the temporary ordinance, but this time failed to get an injunction. ISKCON then challenged that decision. The cases involving both the original and the temporary ordinances then went to the Ninth District Court of Appeals. In seeking to lift the injunction against the original 1997 ordinance, LAX was joined by airports in San Francisco, Oakland, and San Diego. Those airports filed a brief in support of LAX arguing that airports were not public forums and, further, that security dictated that people move through the airport quickly and not be allowed to loiter.[45]

The appeals court referred the case back to the California Supreme Court requesting a ruling on the constitutionality of the ordinance under the state constitution. The court asked whether or not the airport was a public forum under state law and, if so, did the ordinance violate the state constitution. The Supreme Court of California in ruling on the constitutionality of the law argued that it did not have to come to the determination of whether or not the airport was a public forum in order to unanimously uphold the ordinance as valid under the state constitution, though a majority of the justices accepted the argument that LAX was not a public forum. Instead, the decision emphasized that the ordinance was a valid security measure, arguing that the solicitation of funds was problematic given the hectic and crowded conditions within the airport.[46]

Although several news stories concerning the decision declared the state Supreme Court's action final, the case still had to return to the Ninth Circuit Court of Appeals. It was clear, however, that the prospect for successful appeal of the ruling was limited. While Los Angeles officials said they would not enforce the 1997 ordinance until hearing from the Ninth Circuit court, other airports in California, especially San Francisco's, stated that they would explore their options to restrict solicitation activities in light of the state court's decision. In July 2010, the Ninth Circuit Court of Appeals lifted the injunction again the enforcement of the municipal ordinance.[47]

Conclusion

Beginning in the late 1960s and continuing into the twenty-first century, America's airports became locations where first and fourth amendment rights were tested. Despite challenges, in the wake of waves of hijackings, federal officials encouraged and then required the adoption of airport security measures. The security measures now in place have changed the airport travel experience in the USA. Passengers must arrive an hour or two before flights. Non-passengers are restricted from entering the gate area. Departing passengers say their goodbyes at security, not at the gate. Similarly, arriving passengers are no longer greeted at the gate, but must wait until reaching a point beyond security to meet family and friends. And getting through security not only takes time, it is also intrusive. Shoes, belts, and watches must be removed. One might also be subject to not just a magnetometer but a new full-body scanner. And pat-downs are not uncommon. If air travel once meant freedom, today the exercise of that freedom now involves the acceptance of airport security measures that require limitations of constitutional rights.

Notes

1. A bomb exploded on Pan Am 103, a flight which originated in Frankfurt, Germany, and was enroute to Detroit, Michigan, on December 21, 1988. All 243 passenger and 16 crew members aboard died as did 11 people on the ground in Lockerbie, Scotland. On July 17, 1996, TWA 800 experienced a catastrophic explosion, causing the 747 to break apart and crash shortly after take-off from JFK airport, killing all 230 aboard. Though conspiracy theories abound, the National Transportation Safety Board (NTSB) determined a spark ignited fuel vapors in the plane's center fuel tank causing the explosion.
2. James Ott, "FAA Tightens Airport Security to Counter Sabotage Threats," *Aviation Week & Space Technology* 124 (April 21, 1986): 31–32; _____. "Airlines, Airports Prepare to Maintain High Security Levels Over Long Term," *Aviation Week & Space Technology* 124 (April 28, 1986): 29–30.

3. "Dole Directs FAA to Tighten Airport Security," *Aviation Week & Space Technology* 125 (September 15, 1986): 33.
4. Ott, "FAA Tightens Airport Security,"31; _____., "High Security Over Long Term,"30.
5. "Security Issues, Terrorist Threats Concern Airport Operators More than U.S. Airlines," *Aviation Week & Space Technology*, Vol. 124, No. 21 (May 26, 1986): 31.
6. "Airlines Press FAA for New Security Systems," *Aviation Week & Space Technology* 124 (April 28, 1986): 29; Richard Witkin, "U.S. Will Test Bomb Detector for Air Cargo," *New York* Times, April 18, 1987; James Ott, "Airlines, FAA Bolster Research Efforts to Intensify Detection of Aircraft Bombs," *Aviation Week & Space Technology* 134 (March 25, 1991): 54.
7. John H. Cushman, Jr., "Tougher Airline Security Steps Debated," *New York Times*, January 25, 1989; Joan M. Feldman, "The Dilemma of 'Open Skies,'" *New York Times*, April 2, 1989.
8. John H. Cushman, Jr., "Aviation Official Warn of Delays In Steps to Tighten Flight Security," *New York Times*, February 10, 1989; _____., "British Conclude Cassette Player Held Pan Am Bomb," *New York Times*, February 17, 1989; _____., "Security at Airports Steadily Tightening," *New York Times*, May 14, 1989; _____., "Airlines Ordered to Tighten Security Measures," *New York Times*, June 23, 1989; _____., "F.A.A. Admits Delay in Enforcing Security Rules," *New York Times*, December 19, 1989; Breck W. Henderson, "FAA Stays Undecided on Deploying TNA Amid Conflicting Views and Test Results," *Aviation Week & Space Technology* 134 (March 25, 1991): 55.
9. See Aviation Security Improvement Act of 1990 (P. L. 101–604, Title I: Aviation Security); John H. Cushman, Jr., "New Steps Urged for Air Security," *New York Times*, May 15, 1990; James Ott, "Airlines, FAA Bolster Research Efforts," 54; Breck Henderson, "FAA Stays Undecided on Deploying TNA," 55–56; Christopher P. Fotos, "Aviation Security Act Puts TNA Buy on Hold," *Aviation Week & Space Technology* 134 (March 25, 1991): 57–58.
10. John H. Cushman, Jr., "The Crash of Flight 800: Security; To Troubled Airline Industry, Keeping Up With Strict Safety Rules Is Costly Challenge," *New York Times*, July 21, 1996; Pam Belluck and John Sullivan, "The Fate of Flight 800: The Security Gap: Despite Warnings, Most Air Cargo Is Unscreened," *New York Times*, July 26,

1996; "In Bid for Airline Security, Echoes of Unmet Promises," *New York Times*, August 13, 1996.

11. For the official NTSB report on TWA 800 see http://www.ntsb.gov/doclib/reports/2000/AAR0003.pdf (accessed August 27, 2014). Despite the official report, conspiracy theories abound, so many that TWA 800 conspiracy theories have their own Wikipedia page: http://en.wikipedia.org/wiki/TWA_Flight_800_conspiracy_theories (accessed August 27, 2014).
12. Matthew L. Wald, "Airlines and Other Question Aspects of Anti-Terrorism Plan," *New York Times*, September 12, 1996; Todd S. Purdum, "Clinton Signs a Wide-Ranging Measure on Airport Security," *New York Times*, October 10, 1996.
13. Purdum, "Clinton Signs Measure," 32–34.
14. James Ott, "Sober View Takes Over: What Can be Done?" *Aviation Week & Space Technology* 145 (October 7, 1996): 34–35.
15. Carole A. Shifrin, "Airports on Track in Deploying New-Technology Security Devises," *Aviation Week & Space Technology* 149 (September 7, 1998): 166, 169.
16. Matthew L Wald, "Bomb Scanner for Airports is Underused, Audit Finds," *New York Times*, May 16, 1998; _____., "Costly Machines to Check Bags Mostly Idle, Report Says," *New York Times*, October 11, 1998.
17. "Airport Searches and the Fourth Amendment," *Columbia Law Review* 71 (Jun. 1971): 1040–1041.
18. Ibid., 1041–1047.
19. Ibid., 1047–1052: Kraus, 393–394; "The Constitutionality of Airport Searches," *Michigan Law Review* 72 (November 1973): 143.
20. "Airport Searches and the Fourth Amendment," 1052–1053.
21. Ibid., 1052–1054; Kraus, 394–96.
22. "Airport Searches and the Fourth Amendment," 1054–1058.
23. Kraus, 402–411; see also "The Constitutionality of Airport Searches,"128–157.
24. Preston, 42; Wayne R. LaFave, *Search and Seizure: A Treatise on the Fourth Amendment*, Fourth Edition, West's Criminal Practice Series, Vol. 5 (New York: Thompson Reuters Westlaw, 2009), 10.6(c).
25. For examples see Jeffrey A. Carter, "Fourth Amendment. Airport Searches and Seizures: Where will the Court Land?" *The Journal of Criminal Law and Criminology* (*1973-*) 71 (Winter, 1980): 499–517; John M. Schohl, "Airport Seizures off Luggage without Probable

Cause: Are They 'Reasonable'?" *Duke Law Journal* 1982 (Dec. 1982): 1089–1109; Linda M. Sickman, "Fourth Amendment: Limited Luggage Seizures Valid on Reasonable Suspicion," *The Journal of Criminal Law and Criminology* (*1973-*) 74 (Winter, 1983): 1225–1248.

26. *The 9/11 Commission Report*: *The Final Report of the National Commission on Terrorist Attacks upon the United States, Authorized Edition* (New York: W.W. Norton, 2004), 1–14.
27. *The 9/11 Commission Report*, 84–85; "Attacks Highlight Faults in Airport Security," *Aviation Week & Space Technology* 155 (September 17, 2001): 55; John Croft, "Screener Workforce Faces Federal Scrutiny," *Aviation Week & Space Technology* 155 (October 22, 2001): 60–61.
28. David Bond, "Airport Screening Issue Divides House Leadership," *Aviation Week & Space Technology* 155 (October 22, 2001): 58–59.
29. John Croft, "Air Security Bill Clears Lawmakers' Logjam," *Aviation Week & Space Technology* 155 (November 19, 2001): 46.
30. http://www.tsa.gov/stakeholders/screening-partnership-program (accessed August 23, 2014).
31. "TSA: Our History" http://www.tsa.gov/research/tribute/history.shtm (accessed July 9, 2010).
32. Lillian R. BeVier, "Rehabilitating Public Forum Doctrine: In Defense of Categories," *The Supreme Court Review* 1992 (1992): 81–84. See also Harry Kalven, Jr., "The Concept of the Public Forum: Cox v. Louisiana," *The Supreme Court Review* 1965 (1965): 1–32.
33. BeVier, "Rehabilitating Public Forum Doctrine," 84–96.
34. Rhyne, *Airports and the Law*, 23–25.
35. Kathleen M. Sullivan and Akhil Reed Amar, "The Supreme Court, 1991 Term," *Harvard Law Review* 106 (Nov 1991), 279–280.
36. Ibid., 280–281.
37. Ibid., 282.
38. Ibid.
39. For the connection between airports and parks, see Bednarek, "The Flying Machine in the Garden," 354–359.
40. Ibid., 370–372. For examples of the wide range of activities at airports, especially during the 1920s and 1930s, see Ibid., 359, 362–364, 367–370.
41. Will Buchanan, "Court upholds ban on Hare Krishna soliciting in LAX airport," *The Christian Science Monitor* March 26, 2010, http://

www.csmonitor.com/layout/set/print/content/view/print/290402 (accessed March 26, 2010).

42. Lizabeth Cohen, *A Consumer's Republic: The Politics of Mass Consumption in Postwar America* (New York: Vintage Books, Inc., 2003), 274–278.
43. Board of Airport Commissioners of the City of Los Angeles v. Jews for Jesus, Inc. (No. 86–104). http://www.law.cornell.edu/supct/html/historics/USSC_CR_0482_0569_ZO.html. (Accessed 12 May 2010).
44. Robert Lindsey, "Decision by Court Troubles Airports," *New York Times*, July 22, 1987.
45. Maura Dolan and DanWeikel, "L.A. can bar Hare Krisnas from panhandling at airports, court rules," *The Los Angeles Times*, March 26, 2010, http://www.latimes.com/news/local/la-me-lax-hare-krishnas26-2010mar26,0,4257614.story (accessed 12 May 2010); "Denise Nix, "State Supreme Court to hear Krishna LAX free speech case," DailyBreeze.com, January 1, 2010 .http://www.dailybreeze.com/religion/ci_14120775 (accessed 9 March 2010).
46. Dolan and Weikel, "L.A. can bar Hare Krisnas from panhandling at airports, court rules," http://www.latimes.com/news/local/la-me-lax-hare-krishnas26-2010mar26,0,4257614.story (accessed 12 May 2010); "CA Supreme Court rules on Hare Krishnas' Airport Solicitations," http://www.legalinfo.com/legal-news/ca-supreme-court-rules-on-hare-krishnas-airport-solicitations.html (accessed 12 May 2010).
47. Dolan and Weikel, "L.A. can bar Hare Krisnas from panhandling at airports, court rules," http://www.latimes.com/news/local/la-me-lax-hare-krishnas26-2010mar26,0,4257614.story (accessed 12 May 2010); A. Pawlowski, "Panhandling banned at Los Angeles Airport," http://www.cnn.com/2010/TRAVEL/07/08/los.angeles.airport.solicitations/index.html?hpt=Sbin (accessed July 8, 2010).

Conclusion: Airports, Cities, and the Jet Age

In 2013, a group of renown urban specialists from diverse fields, including history, architecture, and American studies, met to "rethink" the American city at a conference in Munich, Germany. Topics discussed included energy, sustainability, aesthetic space, and mobility. In his talk on mobility and the city, Klaus Benesch, a professor of English and American Studies at the Ludwig Maximillian University of Munich, acknowledged the roles the concepts of mobility and "supermodernity" have had in shaping ideas about contemporary cities, particularly their relationship to the "shifting notion of place in modern societies" and the emphasis on "unfettered mobility and excessively mobile lifestyles." However, he also made a case for the continued importance of place, of a reconnection "to the physical environment," especially "spaces of grounding." In doing so, he particularly addressed the notion that airports had become the ultimate example of what Marc Auge termed the "non-place" and John Kasarda's notion that the cities of the future will be aerotropolises, cities centered on their airports, transitional places focused solely on mobility. Particularly in reference to Kasarda, Benesch noted that traditionally cities have always grown around centers. And traditionally "there are different historical layers that are added… but they are all place bound and they somehow are place conscious in terms of where they are." Benesch noted that Kasarda and his model would ignore those layers as "they don't count anymore, because the aerotropolis is just one cog, one wheel, in a huge universal

J.R. Bednarek, *Airports, Cities, and the Jet Age*,
DOI 10.1007/978-3-319-31195-1_9

kind of new city machine or conurbation where the only thing that counts is to be able to have access to a means of transportation that gets you away from the place where you live."[1]

This book is based on the notion that airports are important, real, and historically grounded places in the urban landscape of the USA. They are not ahistorical "non-places." Rather, they are places with multiple layers of history—national and international as well as distinctly local in scope. For example, airports may be locally owned and managed, but they are also parts of a national and international system of transportation. Hence, airport development has involved dialogue and negotiation between local authorities and national authorities. They are also a local place where people and goods find links to nearly everywhere on earth. While highly desired for those very connections, airports are also problematic producers of noise pollution. However, unlike air or water pollution, airport noise pollution effects relatively limited areas. Most people living in cities with even the largest airports can go through their days with little or no bother from jet aircraft noise. Dealing with noise pollution where it is a daily problem, however, has proved a tremendous challenge to local officials and has resulted in significant changes to the landscapes surrounding the nation's airports. Further, airport terminals replaced the train station as the "front door" through which visitors first enter a city. In response, many cities built iconic structures to signal to those visitors the city's civic pride and status. And though airline companies have emphasized the freedom that comes with flight, airports have emerged as places where both first and fourth amendment rights have been challenged and curtailed in the name of security. Thus, airports are the multi-layered products of, among other factors, a particular ownership and management structure, responses to new technologies and an expanding airline travel industry, economic development dreams, and security nightmares.

Airports also have been instrumental in the ability of Americans to take advantage of powered flight. The automobile may be the greatest symbol of Americans' love of freedom and movement, but the airplane has also figured quite prominently. Many early airplane enthusiasts envisioned a future in which the airplane would be as common a form of personal transportation as the automobile. As of the early twenty-first century, we are still waiting for our flying cars. However, Americans have been able to take advantage of the speed afforded by powered flight as passengers on the nation's airlines. Passenger volume has witnessed relatively steady and oftentimes impressive growth from the 1930s through the early twenty-first century. Once the transportation method of choice for just the wealthy or the busy businessman, passengers on the nation's airlines today come from a broad cross-section of the population—male and female, young and old. Recent studies indicated that 42 % of the US adult population reported having flown for leisure in the past year (2008–2009) with that percentage rising to 48 % if travel for business was included.[2] The passenger airline industry has relied on the willingness of local officials to build, manage, and maintain the nation's commercial airports. Though most Americans would not want to have an airport in their backyard, millions use their local, commercial airports every year for business, leisure, and personal travel needs.

And cities have used—or attempted to use—airports as tools in interurban competition. Just as urban boosters scrambled for railroad connections in the nineteenth century, local civic leaders in the twentieth century promoted airport development. They did not believe that any city could survive if it did not have a place on the air transportation map of the USA. With the advent of jet airliners, cities wanted to join the jet age. However, the competitive environment has proved dynamic and often difficult to navigate. Through the late 1970s, government regulation helped create fairly stable and understandable rules of competition. While not allowing everyone to hit the jackpot, it generally limited losses. After 1978, deregulation worked to magnify both the wins and the losses. Deregulation, for example, encouraged airlines to adopt a hub-and-spoke route structure. In the late 1970s and early 1980s, there were more than a dozen major airlines in the USA, and most consolidated many of their flights at select hub airports. Cities competed with one another to gain hub status. For the winners, it meant significant expansion of their airports. However, deregulation also brought airline bankruptcies and mergers. As the number of major airlines declined, so too did the need for multiple hub airports as the newly merged airlines further consolidated their flights. The airports at Atlanta, Chicago, Los Angeles, and Dallas-Fort Worth have emerged as the major winners—though their victories have also challenged local officials to keep up with expanding air traffic. Airport officials in St. Louis and Cincinnati, on the other hand, have the challenge of managing decline. Today, corporate decisions based on the bottom line, not an ideal vision of a national air transportation system, largely determine the fate of airport development projects.

Though an extremely important part of the nation's transportation infrastructure, the economic development toolbox of cities, and the urban landscape, airports remain a little examined subject. While this study offers a broad overview of the development of American airports since 1945, more local studies are necessary. Just as urban historians have gained insight into the complex and often varied histories of such subjects as public housing, urban renewal, highway construction and a myriad of others through detailed case studies, more could and should be learned about the relationship between cities and their airports. How and to what extent do airports reflect local conditions? What is the role of the local airport in economic development? How have local officials responded to the noise pollution generated by jet airliners? What has worked and what has not? What roles have race, class, and gender issues played in airport development – from where airports were built to who finds employment? The field is wide open.

Notes

1. See Klaus Benesch, "Mobility," in Miles Orvell and Klaus Benesch, eds., *Rethinking the American City: An International Dialogue* (Philadelphia: University of Pennsylvania Press, 2014), 143–165.
2. US Travel Association, "Travel Facts and Statistics," http://www.ustravel.org/news/press-kit/travel-facts-and-statistics (access August 1, 2013).

References

Archival and Library Sources

Airports Council-International, Washington, D.C.
Organizational Archives: Minutes of Board Meetings
Cleveland Public Library
Vertical Files, Public Administration
Minnesota State Historical Society, St. Paul, MN
Metropolitan Airports Commission, Subject Files
National Archives-College Park
RG 237: Records of the Federal Aviation Administration, 1922–1981
Omaha Public Library
Clip Files
Washington University, University Archives, St. Louis, MO
Office of the Mayor, Files of Alfonso J. Cervantes, Series Two
Office of the Mayor, Files of Raymond R. Tucker, Series One
Western Historical Manuscript Collection, Thomas Jefferson Library, University of Missouri-St. Louis, St. Louis, MO
Papers of Mayor Vincent Schoemehl, Third Term Records, 1990–1993
Robert Young Papers

Newspapers

Cincinnati Enquirer
Cincinnati Post
Cincinnati Post-Times

J.R. Bednarek, *Airports, Cities, and the Jet Age*,
DOI 10.1007/978-3-319-31195-1

Cincinnati Post-Times-Star
Cleveland Plain Dealer
Cleveland Press
Dayton Daily News
Miami News
Milwaukee Business Journal
Minneapolis Star
Minneapolis Tribune
The Mooresville/Decatur Times
New York Times
Omaha World Herald
Pittsburgh Press
USA Today
Wall Street Journal

Government Documents

Advisory Committee on Intergovernmental Relations. *State Involvement with Federal Local Grant Programs: A Case Study of the "Buying In" Approach.* Washington, D.C.: Advisory Committee on Intergovernmental Relations, December 1970.

Airport and Airway Safety and Capacity Expansion Act of 1987 (P.L. 100–223).

Airport Noise and Capacity Act of 1990, Subtitle D of Title IX of the Omnibus Budget Reconciliation Act of 1990.

Airport Planning Branch, Airports Division, Bureau of Facilities and Material. *Aircraft Noise Abatement.* Washington, D.C.: Federal Aviation Agency, Planning Series, Item No. 3, Sept. 2, 1960.

The Airport and Its Neighbors: The Report of the President's Airport Commission. Washington, D.C.: U.S. Government Printing Office, May 16, 1952.

Aviation Security Improvement Act of 1990 (P. L. 101–604), Title I: Aviation Security.

Committee on Interstate and Foreign Commerce, *Hearings Before the Committee on Interstate and Foreign Commerce, House of Representatives, Seventy-Ninth Congress, First Session on H.R. 3170: A Bill to Provide Aid for the Development of Public Airports and to Amend Existing Laws Relating to Air Navigation Facilities, May 15, 16, 22, 23, 24, 25 29, 30, 31 and June 1 and 5, 1945.* Washington D.C.: Government Printing Office, 1945.

Eilers, Sarah. "Airport Growth: Creating New Economic Development Opportunities." Washington, D.C.: National Council for Urban Economic Development, December 1989.

Federal Aviation Administration, Department of Transportation. "Calendar Year 2005 Primary and Non-Primary Commercial Service Airports." October 31, 2006.

Federal Aviation Administration, Department of Transportation, "Changes in Revenue Passenger Enplanements at Primary Airports Base Year CY 1994 (By Rank Order CY 1995)." October 16, 1996.

Federal Aviation Administration, Department of Transportation. "Enplanements at Primary Airports (Rank Order) CY 10." October 25, 2011.

Federal Aviation Administration, Department of Transportation. "Primary Airport Enplanement Activity Summary for CY 2000 Listed by Rank Order, Enplanements." October 19, 2001.

Federal Aviation Administration, Department of Transportation. "U.S. Airport Enplanement Activity Summary for CY 1985 Listed by Rank Order, Enplanements." 30 September 1986.

Federal Aviation Agency, *The Sounds of the Twentieth Century*. Washington, D.C.: Government Printing Office, 1961.

General Accounting Office. "Report to the Subcommittee on Aviation, Committee on Transportation and Infrastructure, House of Representatives, Airport Privatization: Issues Related to the Sale or Lease of U.S. Commercial Airports." November 1996.

Henry, Eric. *Excise Taxes and the Airport and Airway Trust Fund, 1970–2002.* Washington, D.C.: Statistics of Income Bulletin, 2003.

Kincaid, Ian, Michael Tretheway, Stephanie Gros and David Lews. *Addressing Uncertainty about Future Airport Activity Levels in Airport Decision Making.* Washington D.C.: Transportation Research Board, Airport Cooperative Research Program Report 76, 2012.

Kirk, Robert S. "Airport Improvement Program Reauthorization Legislation in the 106th Congress." Congressional Research Service, The Library of Congress, April 17, 2000.

Metropolitan Council. *Aviation Policy Plan.* Metropolitan Council, adopted December 19, 1996.

Office of Science and Technology, Executive Office of the President. *Alleviation of Jet Aircraft Noise Near Airports: A Report of the Jet Aircraft Noise Panel.* Washington, D.C., March 1966.

Skramstad, Scott. "History of Airport Noise Abatement." Metropolitan Airports Commission, June 2002.

A Staff Report on Federal Air to Airports. Submitted to the Commissions on Intergovernmental Relations, June 1955.

Surface Transportation Assistance Act of 1982 (P.L. 97–249).

Tax Equity and Fiscal Responsibility Act (P.L. 97–248), Title V: Airport and Airways Improvement Act of 1982.

United States v. Causby 328 U.S. 256 (1946).

United State Supreme Court, 369 U.S. 84 *Griggs v. County of Allegheny, Pennsylvania.*

Washington Metropolitan Airports Act of 1986, Title VI of Public Law 99–500.

Books

Ashford, Norma, H. P. Martin Stanton, Clifton A. Moore, *Airport Operations*, 2nd. Ed. New York: McGraw-Hill, 1996.

Altshuler, Alan and David Luberoff, *Mega-Projects: The Changing Politics of Urban Public Investment*. Cambridge, MA, and Washington, D.C.: Lincoln Institute of Land Policy and Brookings Institution Press, 2003.

Anderson, John D. *The Airplane: A History of Technology*. Reston, VA: American Institute of Aeronautics and Astronautics, 2002.

Auge, Marc. *Non-Places: An Introduction to Supermodernity*. London: Verso, second English Language Translation, 2008.

Baron, Robert Alex. *The Tyranny of Noise*. New York: St. Martin's Press, 1970.

Bednarek, Janet R. Daly. *America's Airports: Airfield Development, 1918–1947*. College Station: Texas A&M University Press, 2001.

Berland, Theodore. *The Fight for Quiet*. Englewood, NJ: Prentice-Hall, 1970.

Biles, Roger. *The Fate of Cities: Urban America and the Federal Government, 1945–2000*. Lawrence: The University Press of Kansas, 2011.

Bilstein, Roger E. *The American Aerospace Industry: From Workshop to Global Enterprise*. New York and London: Twayne Publishers, an imprint of Simon & Schuster McMillan and Prentice Hall International, 1996.

Braden, Betsey and Paul Hagen. *A Dream Takes Flight: Hartsfield Atlanta International Airport and Aviation in Atlanta*. Atlanta: Atlanta Historical Scoeity; Athens and London: University of Georgia Press, 1989.

Bragdon, Clifford. *Noise Pollution: The Unquiet Crisis*. Philadelphia: University of Pennsylvania Press, 1970.

Bubb, Daniel K. *Landing in Las Vegas: Commercial Aviation and the Making of a Tourist City*. Reno and Las Vegas: University of Nevada Press, 2012.

Cohen, Lizabeth. *A Consumer's Republic: The Politics of Mass Consumption in Postwar America*. New York: Vintage Books, 2003.

Conway, Erik M. *High Speed-Dreams: NASA and the Technopolitics of Supersonic Transportation, 1945–1999*. Baltimore: The Johns Hopkins University Press, 2005.

Cooper, Phillip J. *The War Against Regulation: From Jimmy Carter to George W. Bush*. Lawrence: University of Kansas Press, 2009.

Corn, Joseph J. *The Winged Gospel: America's Romance with Aviation, 1900–1950*. Oxford and New York: Oxford University Press, 1983.

Courtwright, David T. *Sky as Frontier: Adventure, Aviation, and Empire*. College Station: Texas A&M University Press, 2005.

Dahlstrom, Harl. *A.V. Sorenson and the New Omaha*. Omaha, NE: Lamplighter Press, Douglas County Historical Society, 1987.

Daly-Bednarek, Janet R. *The Changing Image of the City: Planning for Downtown Omaha, 1945–1972*. Lincoln: The University of Nebraska Press, 1992.

Danielson, Michael N. and Jameson W. Doig. *New York: The Politics of Urban Regional Development*. Berkeley: University of California Press, 1982.

Davis, Benjamin O., Jr. *Benjamin O. Davis, Jr.: American, An Autobiography*. Washington, D.C.: Smithsonian Institution Press, 1991.

Depmpsey, Paul Stephen and Andrew R. Goetz. *Airline Deregulation and Laissez-Faire Mythology*. Westport, CT: Quorum Books, 1992).

Demsey, Paul Stephen, Andrew R. Goetz, and Joseph S. Szyliowicz. *Denver International Airport: Lessons Learned*. New York: McGraw Hill, 1997.

Dierikx, Marc and Bram Bouwens. *Building Castles of the Air: Schiphol Amsterdam and the development of airport infrastructure in Europe, 1916–1996*. The Hague: SDU Publishers, 1997.

Doig, Jameson W. *Empire on the Hudson: Entrepreneurial Vision and Political Power at the Port of New York Authority*. New York: Columbia University Press, 2001.

Durkin Keating, Ann. *Chicagoland: Cities and Suburbs in the Railroad Age*. Chicago: University of Chicago Press, 2005.

Erie, Steven P. *Globalizing L.A.: Trade, Infrastructure and Regional Development*. Stanford, CA: Stanford University Press, 2004.

Foster, Mark. *From Streetcar to Superhighway: American City Planners and Urban Transportation*. Philadelphia: Temple University Press, 1981.

Glaab, Charles N. *Kansas City and the Railroads: Community Policy in the Growth of a Regional Metropolis*. Lawrence: University of Kansas Press, 1993.

Gomez-Ibanez, Jose A. and John R. Meyer, *Going Private: The International Experience with Transport Privatization*. Washington, D.C.: The Brookings Institution Press, 1993.

Gordon, Alastair. *Naked Airport: A Cultural History of the World's Most Revolutionary Structure*. New York: Metropolitan Books, 2004.

Hansen, James. *The Bird is on the Wing: Aerodynamics and the Progress of the American Airplane*. College Station: Texas A&M University Press, 2004.

Hardaway, Robert M. *Airport Regulation, Law, and Public Policy*. New York: Quorum Books, 1991.

Hodos, Jerome I. *Second Cities: Globalization and Local Politics in Manchester and Philadelphia*. Philadelphia: Temple University Press, 2011.

Kasarda, John D. and Greg Lindsay. *Aerotropolis: The Way We'll Live Next*. New York: Farrar, Straus and Giroux, 2011.

Kavaler, Lucy. *Noise: The New Menace*. New York: The John Day Company, 1975.

Kent, Richard J., Jr. *Safe, Separated, and Soaring: A History of Federal Civil Aviation Policy, 1961–1972*. Washington, D.C.: US Department of Transportation, Federal Aviation Administration, 1980.

Knauth, Arnold, et. al., eds. *U.S. Aviation Reports, 1928*. Baltimore, MD: United States Aviation Reports, 1928.

Koerner, Brendan. *The Skies Belong to Us: Love and Terror in the Golden Age of Hijacking*. New York: Random House, 2013.

LaFave, Wayne R. *Search and Seizure: A Treatise on the Fourth Amendment, Fourth Edition, West's Criminal Practice Series, Vol. 5*. New York: Thompson Reuters Westlaw, 2009.

Ling, Peter J. *America and the Automobile: Technology, Reform and Social Change*. Manchester: Manchester University Press, 1990.

Lipscomb, David M. *Noise: The Unwanted Sounds*. Chicago: Nelson-Hall Company, 1974.

MacKenzie, David. *IACO: A History of the International Civil Aviation Organization*. Toronto: University of Toronto Press, 2010.

McCartin, Joseph A. *Collision Course: Ronald Reagan, the Air Traffic Controllers, and the Strike that Changed America*. New York: Oxford University Press, 2011.

McShane, Clay. *Down the Asphalt Path: The Automobile and the American City*. New York: Columbia University Press, 1994.

Merket, Jayne. *Eero Saarinen*. New York and London: Phaidon Press, 2005.

Monkkonen, Eric. *The Local State: Public Money and American Cities*. Stanford, CA: Stanford University Press, 1985.

Nelkin, Dorothy. *Jetport: The Boston Airport Controversy*. New Brunswick, NJ: Transaction Books, 1974.

The 9/11 Commission Report: The Final Report of the National Commission on Terrorist Attacks upon the United States, Authorized Edition. New York: W.W. Norton, 2004.

Page, Max. *The Creative Destruction of Manhattan, 1900–1940*. Chicago: The University of Chicago Press, 1999.

Pearman, Hugh. *Airports: A Century of Architecture*. New York: Harry N. Abrams, Inc., 2004.

Preston, Edmund. *Troubled Passage: The Federal Aviation Administration During the Nixon-Ford Term, 1973–1977*. Washington, D.C.: US Department of Transportation, Federal Aviation Administration, 1987.

Ragsdale, Kenneth B. *Austin: Cleared for Take-off: Aviators, Businessmen, and the Growth of an American City*. Austin: University of Texas Press, 2004.

Rhyne, Charles. *Airports and the Courts*. Washington, D.C.: National Institute of Municipal Law Officers, 1944.

Rhyne, Charles. *Airports and the Law*. Washington, D.C.: National Institute of Municipal Law Officers, 1979.

Rochester, Stuart I. *Takeoff at Mid-Century: Federal Civil Aviation Policy in the Eisenhower Years, 1953–1961*. Washington, D.C.: US Department of Transportation, Federal Aviation Administration, 1976.

Rose, Mark, Bruce E. Seely, and Paul Barrett. *The Best Transportation System in the World: Railroads, Trucks, Airlines, and American Public Policy in the Twentieth Century*. Columbus: The Ohio University Press, 2006.

Salter, Mark B., ed. *Politics at the Airport*. Minneapolis: University of Minnesota Press, 2008.

Sanders, Haywood T. *Convention Center Follies: Politics, Power and Public Investment in American Cities*. Philadelphia: University of Pennsylvania Press, 2014.

Sbragia, Alberta M. *Debt Wish: Entrepreneurial Cities, U.S. Federalism, and Economic Development*. Pittsburgh, PA: University of Pittsburgh Pres, 1996.

Stevenson, Gordon McKay, Jr. *The Politics of Aircraft Noise*. Belmont, CA: Duxbury Press, 1972.

Urry, John. *Mobilities*. Cambridge: Polity Press, 2007.

Wachs, Martin and Margaret Crawford, eds. *The Car and the City: The Automobile, the Built Environment and Daily Urban Life*. Ann Arbor: University of Michigan Press, 1992.

Warner, Sam Bass, Jr. *Streetcar Suburbs: The Process of Growth in Boston (1870–1900)*. Boston, MA: Harvard University Press, 1962.

Wilson, John R. M. *Turbulence Aloft: The Civil Aeronautics Administration Amid Wars and Rumors of Wars, 1938–1953*. Washington, D.C.: US Department of Transportation, Federal Aviation Administration, 1979.

Winston, Clifford. *Last Exit: Privatization and Deregulation of the U.S. Transportation System*. Washington, D.C.: The Brookings Institution Press, 2010.

Yamasaki, Minoru. *A Life in Architecture*. New York: Weatherhill, Inc., 1979.

Zukowsky, John, ed. *Building for Air Travel: Architecture and Design for Commercial Aviation*. Munich and Chicago: Prestl and The Art Institute of Chicago, 1996.

Articles and Chapters

"Accord Reached by Conferees on Bill for Airport, Airway Aid." *Aviation Week & Space Technology* 104 (June 7, 1976).

"Aircraft Noise: Transport design is key to problem, PNYA says; Official warns of need for better equipment." *Aviation Week* 60 (March 8, 1954).

"Airline, Airport Security Drive Still Faces Financial Obstacles." *Aviation Week & Space Technology* 96 (March 20, 1972).

"Airlines Press FAA for New Security Systems." *Aviation Week & Space Technology* 124 (April 28, 1986).

"Airport Aid." *Aviation Week* 60 (April 26, 1954).

"Airport Aid." *Aviation Week* 61 (October 1954).

"Airport Aid Allocations Cut $29 Million." *Aviation Week & Space Technology* 85 (December 26, 1966).

"Airport Aid Funding Resumed After Year." *Aviation Week & Space Technology* 105 (August 30, 1976).

"Airport, Airways Cost Scrutinized." *Aviation Week & Space Technology* 98 (January 15, 1973).

"Airport, Airways Fund Plans Hit." *Aviation Week & Space Technology* 88 (May 27, 1968).

"Airport Chiefs Urged to Prepare For Airline Concorde Operations." *Aviation Week & Space Technology* 200 (July 1, 1974).

"Airport Cities: Gateways to the Jet Age." *Time Magazine* (August 15, 1960).

"Airport Fund Delay Draws Opposition." *Aviation Week & Space Technology* 113 (August 18, 1980).

"Airport Funds." *Aviation Week* 60 (May 17, 1954).

"Airport-Less Towns in Line for U.S. Aid Under New CAA Policy." *Business Week,* (October 22, 1955).

"Airport Noise Proposals Readied." *Aviation Week & Space Technology* 103 (October 27, 1975).

"Airport Operator Noise Plans, Surcharge on Tickets Proposed." *Aviation Week & Space Technology* 106 (March 14, 1977).

"Airport Operators Seeking Cannon Bill Modifications." *Aviation Week & Space Technology* 111 (December 3, 1979).

"Airport Renovation, Expansion to Increase in 1985." *Aviation Week & Space Technology* 121 (November 12, 1984).

"Airports Devise Economic Survival Plans" *Aviation Week & Space Technology* 115 (November 9, 1981).

"Airport Security Measures Studied." *Aviation Week & Space Technology* 93 (November 9, 1970).

"Airports Seek to Shift Security Burden." *Aviation Week & Space Technology* 98 (January 22, 1973).

"Airport Searches and the Fourth Amendment." *Columbia Law Review* 71 (Jun. 1971).

"Air Power Boosted by Increasing Airports." *Aviation Week* 52 (February 27, 1950).

"Air safety's unspent billions." *Business Week* (February 20, 1978).

"Attacks Highlight Faults in Airport Security." *Aviation Week & Space Technology* 155 (September 17, 2001).

"Aviation: Jets Across the U.S." *Time Magazine* LXXII (November 17, 1958).

"A Bad Case of Vetophobia." *The Nation* 188 (June 27, 1959).

Barrett, Paul. "Cities and Their Airports: Policy Formation, 1926–1952." *Journal of Urban History* 14 (Nov., 1987).

Bassett, Edward W. "Kahn Challenges Aid on Noise." *Aviation Week & Space Technology* 108 (June 19, 1978).

Baxter, William F. and Lillian R. Altree, "Legal Aspects of Airport Noise," *Journal of Law and Economics* 15 (April 1972).

Becker, William B. "Aircraft Noise and the Airlines." *SAE Transactions* 83 (1972).

Bednarek, Janet R. Daly. "The Flying Machine in the Garden: Parks and Airports, 1918–1938." *Technology and Culture* 46 (April 2005).

Bednarek, Janet R. Daly. "Layer Upon Layer: Public Authorities and Airport Ownership and Management in St. Louis, 1947–1980." *Journal of Planning History* 8 (November 2009).

Benesch, Klaus. "Mobility," in Miles Orvell and Klaus Benesch, eds., *Rethinking the American City: An International Dialogue* (Philadelphia: University of Pennsylvania Press, 2014).

BeVier, Lillian R. "Rehabilitating Public Forum Doctrine: In Defense of Categories." *The Supreme Court Review* 1992.

"Bill Would Alter Airport Funding Method." *Aviation Week and Space Technology* 111 (September 3, 1979).

"Bill Would Drop Grants to 69 Airports." *Aviation Week & Space Technology* 114 (March 23, 1981).

Bond, David. "Airport Screening Issue Divides House Leadership." *Aviation Week & Space Technology* 155 (October 22, 2001).

Brodherson, David. "An Airport in Every City: The History of American Airport Design," in John Zukowski, ed., *Building for Air Travel: Architecture and Design for Commercial Aviation*. Munich and New York: Prestel-Verlag; Chicago: The Art Institute of Chicago, 1996.

Bruegmann, Robert. "Airport City," in John Zukowski, ed., *Building for Air Travel: Architecture and Design for Commercial Aviation*. Munich and New York: Prestel-Verlag; Chicago: The Art Institute of Chicago, 1996.

Bulban, Erwin J. "Airport Officials Hit FAA, DOT on Noise, Fund Issues." *Aviation Week & Space Technology* 99 (October 29, 1973).

Bulban, Erwin J. "U.S. Mapping Program to Alleviate Noise." *Aviation Week & Space Technology* 85 (October 24, 1966).

"CAA Issues First Study Report On '100 Problems' of Jet Age." *Aviation Week* 65 (July 16, 1956).

"CAA Liberalizes Airport Aid Policy." *Aviation Week & Space Technology* 63 (October 17, 1955).

"Cannon Willing to Discuss Head Tax in Airport Bill." *Aviation Week & Space Technology* 112 (January 21, 1980).

Carter, Jeffrey A. "Fourth Amendment. Airport Searches and Seizures: Where will the Court Land?" *The Journal of Criminal Law and Criminology (1973-)* 71 (Winter, 1980).

Carter, Joseph W. "N. Y. Jetport Still Becalmed Amid Debate." *Aviation Week & Space Technology* 90 (May 19, 1969).

Carter, Joseph W. "N. Y. "10 Carriers' New York Area Plan Discounts Need for New Jetport." *Aviation Week* 83 (November 1, 1966).

"Civil Aviation Faces Budget Slash in '55." *Aviation Week* 59 (December 14, 1953).

"Chicago's Midway Nation's Busiest Field. " *Aviation Week* 62 (April 11, 1955).

Christian, George L. "Scientists Tackle Jet Noise Problem." *Aviation Week* 65 (September 3, 1956).

Colclough, William G., Lawrence A. Daellenbach, and Keith R. Sherony, "Estimating the Economic Impact of a Minor League Baseball Stadium.," *Managerial and Decision Economics* 15 (Sept-Oct 1994).

"A collision course on financing airports," *Business Week* (May 18, 1981).

"Comment: The Constitutionality of Local Anti-Pollution Ordinances." *Fordham Urban Law Journal* 1(1972).

"Commercial Jet Outlook Bright for U.S. Airways and Airports." *Aviation Week* 68 (March 3, 1958).

"Committee Boosts Airport Aid Funds." *Aviation Week* 58 (June 8, 1953).

"Concorde Ban Termed 'Discriminatory'." *Aviation Week & Space Technology* 106 (June 13, 1977).

"Concorde Stirs Massport Dissension." *Aviation Week & Space Technology* 100 (June 24, 1974).

"Concorde Suit In New York Called Pivotal." *Aviation Week & Space Technology* 104 (May 24, 1976).

"Congress Fights Airport Cuts." *Aviation Week* 58 (May 4, 1953).

"The Congress: Score One for Persistence." *Aviation Week & Space Technology*, Vol. 96, No. 24 (December 14, 1970).

"Congress Unit to Urge More Aid for Development of Civil Airports." *Aviation Week & Space Technology* 86 (January 2, 1967).

"Congress Votes Aircraft Noise Rule Delays." *Aviation Week & Space Technology* 112 (February 11, 1980).

"Congress Weighs Airline Taxes." *Aviation Week & Space Technology* 146 (February 17, 1997).

"The Constitutionality of Airport Searches." *Michigan Law Review* 72 (November 1973).

Cook, Robert H. "Atlanta, Miami Share Concepts, Problems." *Aviation Week & Space Technology* 79 (August 5, 1963).

Cook, Robert H. "CAB Regional Airport Plan Faces Delay." *Aviation Week & Space Technology* 80 (April 20, 1964).

Cook, Robert H. "Idlewild Sprawl Poses Transfer Problems." *Aviation Week & Space Technology*, 79 (July 29, 1963).

Cook, Robert H. "O'Hare Walking Distance Stirs Criticism." *Aviation Week & Space Technology*, 79 (July 15, 1963).

Cook, Robert H. "Pyle Says Jet Noise Still Major Problem." *Aviation Week* 69 (July 28, 1958).

Cook, Robert H. "San Francisco Airport Expansion Started." *Aviation Week & Space Technology* 78 (June 24, 1963).

Croft, John. "Air Security Bill Clears Lawmakers' Logjam." *Aviation Week & Space Technology* 155 (November 19, 2001).

Croft, John. "Screener Workforce Faces Federal Scrutiny." *Aviation Week & Space Technology* 155 (October 22, 2001).
"Delay of Noise Deadline Studied." *Aviation Week & Space Technology* 110 (April 9, 1979).
"Delta Wants to Resolve Idlewild's Jet-Noise Complaint Out of Court." *Aviation Week*, 73 (November 7, 1960).
Dieterich-Ward, Allen. "From Satellite City to Burb of the 'Burgh: Deindustrialization and Community Identity in Steubenville, Ohio," in James J. Connolly, ed. *After the Factory: Reinventing America's Industrial Small Cities.* Lanham, MD: Lexington Books, 2010.
"Dole Directs FAA to Tighten Airport Security." *Aviation Week & Space Technology* 125 (September 15, 1986).
"DOT Proposes New Trust Fund To Aid Noise Curb Compliance." *Aviation Week & Space Technology* 106 (January 24, 1977).
Douglas, Deborah G., "Airports as Systems and Systems of Airports: Airports and Urban Development in America before World War II," in William Leary, ed., *From Airships to Airbus: The History of Civil and Commercial Aviation, Volume I: Infrastructure and Environment.* Washington, D.C.: Smithsonian Institution Press, 1996.
Dunham, Allison. "Griggs v. Allegheny County in Perspective: Thirty Years of Supreme Court Expropriation Law." *The Supreme Court Review*, 1962.
Dyble, Louise Nelson. "The Defeat of the Golden Gate Authority: A Special District, a Council of Governments, and the Fate of Regional Planning in the San Francisco Bay Area." *The Journal of Urban History* 34 (January 2008).
Eastman, Ford. "Airlines Say Jet Noise to Be Cut To Reasonable Level by 1959." *Aviation Week* 66 (May 13, 1957).
Eastman, Ford. "CAA Says Jet Airliner Problems Ease." *Aviation Week* 67 (August 19, 1957).
Eastman, Ford. "Needs of Airports for Jets Stir Debate." *Aviation Week* 70 (March 9, 1959).
"Editorial: Airports and the Trust Fund." *Aviation Week and Space Technology* 111 (November 19, 1979).
"Editorial: The Noise Fraud." *Aviation Week & Space Technology* 109 (August 28, 1978).
"Editorial: Trusting the Trust Fund." *Aviation Week & Space Technology* 111 (October 22, 1979).
Ellingsworth, Rosalind K. "Ford Team Readies Final Reform Try." *Aviation Week & Space Technology* 105 (November 15, 1976).
Ellingsworth, Rosalind K. "Further Cut in Noise Standards Sought. *Aviation Week & Space Technology* 105 (October 11, 1976).
Ellingsworth, Rosalind K. "House Unit Eases Noise Rule Deadline." *Aviation Week & Space Technology* 107 (July 4, 1977).

Ellingsworth, Rosalind K. "Noise Abatement Fund Studied." *Aviation Week & Space Technology* 104 (March 29, 1976).

Evans, Alona E. "Aircraft Hijacking: What is Being Done?" *The American Journal of International Law* 67(October 1973).

"Extended 707 Tests Ease Dispute." *Aviation Week* 69 (September 15, 1958).

"FAA Anti-Noise Authority Seen Clearing House Unit." *Aviation Week & Space Technology* 88 (March 25, 1968).

"FAA Hit by House Unit for Lag in Airports/Airways Obligations." *Aviation Week & Space Technology* 96 (May 29, 1972).

"FAA Noise Position." *Aviation Week* 71 (September 21, 1959).

"FAA Threatens Action Against Airport." *Aviation Week & Space Technology* 112 (April 14, 1980).

"FAA Trying Again to Top Trust Fund." *Aviation Week & Space Technology* 102 (February 17, 1975).

Fairbanks, Robert. "A Clash of Priorities: The Federal Government and Dallas Airport Development," in Joseph F. Rishel, ed., *American Cities and Towns: Historical Perspectives*. Pittsburgh, PA: Duquesne University Press, 1992.

Fairbanks, Robert. "Responding to the Airplane: Urban Rivalry, Metropolitan Regionalism, and Airport Development in Dallas, 1927–1965," in Hamilton Cravens, Alan Marcus, and David M. Katzman, eds. *Technical Knowledge in American Culture: Science, Technology and Medicine Since the Early 1800s*. Tuscaloosa: The University of Alabama Press, 1996.

Feazel, Michael. "Airport Aid Delay Until 1981 Expected." *Aviation Week & Space Technology* 113 (October 13, 1980).

"Fiscal, Social Obstacles Slow Airport Advances." *Aviation Week & Space Technology* 95 (November 15, 1971).

Fotos, Christopher P. "Aviation Security Act Puts TNA Buy on Hold." 134 *Aviation Week & Space Technology* 134 (March 25, 1991).

Fotos, Christopher P. "New Law to Permit Head Tax, Phase Out Stage 2 Aircraft." *Aviation Week & Space Technology* 133 (November 5, 1990).

Fotos, Christopher P. "Pittsburgh Expanding Capacity with Midfield Terminal Project." *Aviation Week & Space Technology* 130 (June 12, 1989).

Fotos, Christopher P. "Revolutionary Terminal Opens Era in Pittsburgh," *Aviation Week & Space Technology* 137 (October 5, 1992).

Freestone, Robert. "Planning, Sustainability and Airport-led Urban Development." *International Planning Studies* 14 (May 2009).

Friedman, Paul D. "Birth of an Airport: From Mines Field to Los Angeles International, L.A. Celebrates the 50th Anniversary of its Airport." *Journal of American Aviation Historical Society* (Winter 1978).

Friedman, Paul D. "Fear of Flying: Airport Noise, Airport Neighbors." *The Public Historian* 1 (Summer 1979).

Garrison, Glenn. "Airline Costs Soar in New Idlewild Unit." *Aviation Week* 68 (June 23, 1958).

Garrison, Glenn. "CAA Borrows B-47 to Check Jet Problems." *Aviation Week* 64 (April 23, 1956).

Garrison, Glenn. "First Jets Will Use Interim Facilities." *Aviation Week* 68 (March 17, 1958).

Garrison, Glenn. "Fourth N. Y. Airline Airport Disputed." *Aviation Week* 74 (May 1, 1961).

Garrison, Glenn. "Hope for Jet Noise Compromise Shaken." *Aviation Week* 69 (September 8, 1958).

Garrison, Glenn. "Inadequate Facilities Jeopardize Jet Era." *Aviation Week* 67 (December 9, 1957).

Garrison, Glenn. "NATCC's Idelwild Anti-Noise Plan Offers Economical Jet Operations." *Aviation Week* 70 (June 29. 1959).

Garrison, Glenn. "New Jersey Airport Plan Faces Hurdles." *Aviation Week* 71 (December 21, 1959).

Garrison, Glenn. "N. Y. Port Authority Takes Delta To Court Over Idlewild Jet Noise." *Aviation Week* 73 (October 31, 1960).

Garrison, Glenn. "Port Authority, Airlines Analyze FAA Idelwild Jet Anti-Noise Rules." *Aviation Week* 72 (May 16, 1960).

Garrison, Glenn. "Port Authority Breaks with NATCC." *Aviation Week* 71 (September 21, 1959).

Garrison, Glenn. "Port Authority Noise Investigators Criticize American's Jet Takeoffs." *Aviation Week* 70 (March 30, 1959).

Goetz, Andrew R. "Air Passneger Transportation and Growth in the U.S. Urban System, 1950–1987." *Growth and Change* 23 (Spring 1992).

Goetz, Andrew R. Goetz Christopher J. Sutton. "The Geography of Deregulation in the U.S. Airline Industry." *Annals of the Association of American Geographers* 87 (June 1997).

"Gov. Carey Voices Opposition to Concorde." *Aviation Week & Space Technology* 104 (March 1, 1976).

"Grants to Largest Airports Not Necessary, Bond Says." *Aviation Week & Space Technology* 111 (September 17, 1979).

Green, William C. "The War against the States in Aviation." *Virginia Law Review* 31 (Sep. 1945).

Groothuis, Peter, Bruce K. Johnson, and John C. Whitehead, "Public Funding of Professional Sports Stadiums: Public Choice or Civic Pride?" *Eastern Economic Journal* 30 (Fall 2004).

Hanks, Donoh W., Jr., "Neglected Cities Turn to U.S." *National Municipal Review* 35 (Apr., 1946).

Harper, Donald V. "The Dilemma of Aircraft Noise at Major Airports." *Transportation Journal* 10 (Spring 1971).

Harper, Donald V. "The Minneapolis-St. Paul Metropolitan Airports Commission." *Minnesota Law Review* 55 (1971).

Harvey, William B. "Landowners' Rights in the Air Age: The Airport Dilemma." *Michigan Law Review* 56 (June 1958).

"Hearings Begin on Bill To Boost Airport Aid." *Aviation Week* 70 (January 26, 1959).

Henderson, Breck. "FAA Stays Undecided on Deploying TNA Amid Conflicting Views and Test Results." *Aviation Week & Space Technology* 134 (March 25, 1991).

Henig, Jeffrey R. "Privatization in the United States: Theory and Practice." *Political Science Quarterly* 104 (Winter, 1989–1990).

Hieronymus, William S. "New Landing Method Aimed at Reduction in Approach Noise." *Aviation Week & Space Technology* 94 (March 1, 1971).

Hoffman, David A. "Noise Returns as Major Airline Problem." *Aviation Week and Space Technology* 76 (April 16, 1962).

Hotchkiss, Henry G. "Airports Before the Bench." *Aero Digest* 55 (Aug., 1947).

"Hot Weather Adds to Jet Noise Problem." *Aviation Week* 70 (June 22, 1959).

"House Conference Participation Key to Maneuvers on Noise Bill." *Aviation Week & Space Technology* 111 (November 12, 1979).

"House Panel Readies Own Noise Rules." *Aviation Week & Space Technology* 104 (March 1, 1976).

"House, Senate Airport Aid Bills Differ." *Aviation Week & Space Technology* 114 (May 4, 1981).

Hunter, George S." Los Angeles Moves to Meet Traffic Gain." *Aviation Week & Space Technology* 87 (July 17, 1967).

"Idlewild to Get New Concept in Terminals." *Aviation Week* 62 (February 28, 1955); 87–88.

Infanger, John F. "Downtown Rebound." *Airport Business* 18 (October 2004).

"Interest Rates, Economy Deter Airport Financing." *Aviation Week & Space Technology* 115 (November 8, 1981).

Irwin, Michael D. and John D. Kasarda. "Air Passenger Linkages and Employment Growth in U.S. Metropolitan Areas." *American Sociological Review* 56 (August 1991).

Issel, William. "'Land Values, Human Values, and the Preservation of the City's Treasured Appearance': Environmentalism, Politics and the San Francisco Freeway Revolt." *Pacific Historical Review* 68 (November 1999).

"Jet Crash Spurs Pilot Criticism Of Noise Abatement Procedures." *Aviation Week & Space Technology* 76 (March 19, 1962).

Johnsen, Katherine. "Democrats Push Airport Aid Measure." *Aviation Week* 70 (February 2, 1959).

Johnsen, Katherine. "House Opens Fire on Airport Aid." *Aviation Week* 58 (April 20, 1953).

Kahn, Richard S. "Stewart Expansion Stirs Mixed Reaction." *Aviation Week & Space Technology* 95 (July 19, 1971).
Kalven, Harry, Jr. "The Concept of the Public Forum: Cox v. Louisiana." *The Supreme Court Review* 1965.
Karsner, Douglas. "Aviation and Airports: The Impact on the Economic and Geographic Structure of American Cities, 1940s-1980s." *Journal of Urban History* 23 (May 1997).
Kolk, F. W. "The First Year of the Jet Age…Reflections." *SAE Transactions* 68 (1960).
Kozicharow, Eugene. "Illinois to Sue Over Noise Rules." *Aviation Week & Space Technology* 105 (August 2, 1976).
Kraus, Douglas M. "Searching for Hijackers: Constitutionality, Costs, and Alternatives." *The University of Chicago Law Review* 40 (Winter 1973).
"L. A. Airport Expansion Faces Vote." *Aviation Week* 58 (Mar 11, 1953).
"LA Airport Fights For Expansion Funds." *Aviation Week* 60 (February 8, 1954).
"Legislation Gave Cities Broader Airport Control." *The American City* 57 (April 1942).
"Legislative Action on Airport Noise May Follow West Coast Hearings." *Aviation Week & Space Technology* 72 (May 2, 1960).
Lenorovitz, Jeffrey M. "Airport Noise Controls at Issue in Trial." *Aviation Week & Space Technology* 112 (January 21, 1980).
Lenorovitz, Jeffrey M. "FAA Irked by Orange County Action." *Aviation Week & Space Technology* 112 (May 26, 1980).
Lewis, Craig. "Ft. Worth Prepares Airport for Jet Era." *Aviation Week* 67 (October 28, 1957).
"Los Angeles Starts Jet Age Terminal." *Aviation Week* 67 (December 2, 1957).
"Los Angeles Terminal to Expand." *Aviation Week* 67 (November 4, 1957).
"Los Angeles Traffic Outstrips Expansion." *Aviation Week* 63 (December 19, 1965).
Mann, Paul. "FAA Budget Breakthrough Reached Just in Time." *Aviation Week & Space Technology* 152 (March 6, 2000).
Mann, Paul. "FAA's Budget Fate Remains Uncertain." *Aviation Week & Space Technology* 151 (October 11, 1999).
Mayo, Louis H. "Consideration of Environmental Noise Effects in Transportation Planning by Governmental Entities." *SAE Transactions* 83 (1972).
McKenna, James T. "FAA Would Tap Trust Fund to Finance Operations in 2000," *Aviation Week & Space Technology* 150 (February 8, 1999).
McKenna, James T. "Hill Lobbyists See No Quick Fix on FAA Funding." *Aviation Week & Space Technology* 150 (January 25, 1999).
McQuade, Walter. "The Birth of an Airport." *Fortune* LXV (March 1962).
"Multi-City Airports Increase." *The American City* 57 (October 1942).

Mohl, Raymond A. "The Interstates and the Cities: The U.S. Department of Transportation and the Freeway Revolt, 1966–1973." *The Journal of Policy History* 20 (Spring 2008).

Mohl, Raymond A. "Stop the Road: Freeway Revolts in American Cities," *Journal of Urban History* 30 (July 2004).

"New Administration Backs Airport Defederalization." *Aviation Week & Space Technology* 114 (April 6, 1981).

"New Noise Requirements Detailed." *Aviation Week & Space Technology* 90 (November 17, 1969).

"New York Area Noise Rules Vetoed." *Aviation Week & Space Technology* 108 (January 2, 1978).

"New York Ban On Concorde Continued." *Aviation Week & Space Technology* 107 (July 11, 1977).

"Noise Funding Shift to Carriers Urged." *Aviation Week & Space Technology* 110 (February 19, 1979).

"Noise Plan Threatens Older Jets." *Aviation Week & Space Technology* 88 (March 18, 1968).

"Noise Reduction Funding Bill Passes House Easily." *Aviation Week & Space Technology* 109 (September 18, 1978).

Noll, Roger G. and Andrew Zimbalist, "Sports, Jobs, Taxes: Are New Stadiums Worth the Cost?" *The Brookings Review* 15 (Summer 1997).

"Opposition Arises to Airport Plans, Fees," *Aviation Week & Space Technology* 114 (April 6, 1981).

Ott, James. "Airlines, Airports Prepare to Maintain High Security Levels Over Long Term." *Aviation Week & Space Technology* 124 (April 28, 1986).

Ott, James. "Airlines, FAA Bolster Research Efforts to Intensify Detection of Aircraft Bombs." *Aviation Week & Space Technology* 134 (March 25, 1991).

Ott, James. "AIR21 Fails in Congress." *Aviation Week & Space Technology*, 151 (November 29, 1999).

Ott, James. "Bush Order Opens Door for Airport Privatization." *Aviation Week & Space Technology* 136 (May 11, 1992).

Ott, James. "FAA Funding Crisis Bolsters Ticket Tax." *Aviation Week & Space Technology* 146 (February 10, 1997).

Ott, James. "FAA Noise Policy to Stress National System, Fleet Replacement," *Aviation Week & Space Technology* 125 (November 17, 1986).

Ott, James. "FAA Rejects Two Proposals to Privatize Albany Airport." *Aviation Week & Space Technology* 131 (December 11, 1989).

Ott, James. "FAA Tightens Airport Security to Counter Sabotage Threats." *Aviation Week & Space Technology* 124 (April 21, 1986).

Ott, James. "Indianapolis Serves as Privatization Testbed." *Aviation Week & Space Technology* 149 (December 14, 1998).

Ott, James. "Plans to Fund New Airports Stir Battle in Congress." *Aviation Week & Space Technology* 132 (February 12, 1990).
Ott, James. "Sober View Takes Over: What Can be Done?" *Aviation Week & Space Technology* 145 (October 7, 1996).
"Palmdale Airport Recommended," *Aviation Week & Space Technology* 108 (May 8, 1978).
Perkinson, Allan C. "The Proposed National Airport Plan." *Virginia Law Review* 31 (Mar., 1945).
Phillips, Edward H. "Airports, Air Services Targeted for Cuts." *Aviation Week & Space Technology* 142 (February 6, 1995).
Phillips, Edward H. "Congress Clears Path for FAA Reform." *Aviation Week & Space Technology* 143 (October 30, 1995).
Phillips, Edward H. "FAA Reform Bill Faces Fight From General Aviation." *Aviation Week & Space Technology*143 (September 18, 1995).
Phillips, Edward H. "FAA User Fees Cloud Reform Initiative." *Aviation Week & Space Technology* 143 (October 2, 1995).
"Pilots Take Tougher Noise Stand, Define Unacceptable Maneuvers." *Aviation Week and Space Technology* 76 (June 11, 1962).
"Planning the 'aerotropolis'." *Airport World* 5 (Oct-Nov 2000).
Poole, Robert W., Jr. "Privatization: A New Transportation Paradigm." *Annals of the American Academy of Political and Social Sciences* 553 (Sep., 1997).
"Port Authority Defends Idlewild Planning Criticized by N. Y. Mayor." *Aviation Week* 64 (February 6, 1956).
Pritchett, Wendell E., and Mark Rose. "Introduction: Politics and the American City, 1940–1990." *Journal of Urban History* 34 (January 2008).
"Proposed FAA Noise Rules Challenged." *Aviation Week* 73 (August 29, 1960).
Proctor, Paul. "FAA Begins Formal Investigation Of Transport Aircraft Noise." *Aviation Week & Space Technology* 130 (February 6, 1989).
Radford, Gail. "From Municipal Socialism to Public Authorities: Institutional Factors in the Shaping of American Public Enterprise." *The Journal of American History* 90 (December 2003).
Randolph, Anne. "Completion of 1984 Slated for Los Angeles Upgrade." *Aviation Week & Space Technology* 118 (February 28, 1983).
"Real Estate Decline Pegged on Jet Noise." *Aviation Week & Space Technology* 74 (June 12, 1961).
"Record Airport Aid Bill Permits Funds for Terminal Construction." *Aviation Week* 63 (July 25, 1955).
"Reduced-Noise Approach Profiles Tested." *Aviation Week & Space Technology* 95 (September 13, 1971).
Ropelewski, Robert R. "Traffic Growth, Noise Wrack Los Angeles Airport Planning." *Aviation Week & Space Technology* 95 (November 15, 1971).

Roseau, Nathalie. "Reach for the Skies, Aviation and Urban Visions: Paris and New York, c. 1910." *Journal of Transport History* 30 (December 2009).
Rosendahl, C. E. "Aircraft Noise Problem in Airport Vicinities." *SAE Transactions* 65 (1955).
Roth, Gabriel. "Airport Privatization." *Proceedings of the Academy of Political Science* 36 (1987).
Ruppenthal, Karl M., "No Money for Safe Landings." *The Nation* 207 (August 26, 1968).
Santala, Susanna. "Airports," in Eeva-Liisa Pelkonen and Donald Albrecht, eds. *Eero Saarinen: Shaping the Future*. New Haven and London: Yale University Press, 2006.
Schechter, Stephen L. "The Concorde and Port Noise Complaints: The Commerce and Supremacy Clauses Enter the Supersonic Age." *Publius* 8 The State of American Federalism, 1977 (Winter 1978).
Schneider, Charles E. "Fragmented Noise Control Sought." *Aviation Week & Space Technology* 99 (July 9, 1973).
Schneider, Charles E. "Noise Control Proposals Hit Jets." *Aviation Week & Space Technology* 98 (May 14, 1973).
Schneider, Charles E. "Volpe Pledges Aviation Tax Fund Integrity." *Aviation Week & Space Technology* 94 (May 3, 1971).
Schohl, John M. "Airport Seizures off Luggage without Probable Cause: Are They 'Reasonable'?" *Duke Law Journal* 1982 (Dec. 1982).
"Second D.C. Airport Gets New Backing." *Aviation Week* 63 (August 15, 1955).
"Security Issues, Terrorist Threats Concern Airport Operators More than U.S. Airlines." *Aviation Week & Space Technology* 124 (May 26, 1986).
"Senate, House Groups Oppose Airport Bill." *Aviation Week* 61 (July 12, 1954).
"Senate Sponsor Sees Nixon Veto of Trust Fund Use Restriction." *Aviation Week & Space Technology* 95 (October 18, 1971).
"Senator Calls for Private Sale of National, Dulles Airports." *Aviation Week & Space Technology* 124 (April 14, 1986).
Shifrin, Carole A. "Airports on Track in Deploying New-Technology Security Devises." *Aviation Week & Space Technology* 149 (September 7, 1998).
Shifrin, Carole A. "Deregulation Bringing Airports More Interest in Own Destiny." *Aviation Week & Space Technology* 121 (November 12, 1984).
Shifrin, Carole A. "FAA Reforms Set in Motion." *Aviation Week & Space Technology* 144 (April 1, 1996).
Shifrin, Carole A. "Official Hopes Capacity Crisis Will Spur Expansion of Airports," *Aviation Week & Space Technology* 127 (November 8, 1987).
Shifrin, Carole A. "U.S. Nears Transfer of National, Dulles Control." *Aviation Week & Space Technology* 122 (April 1, 1985).
"Shifting Aircraft Noise Liability to the Federal Government." *Virginia Law Review* 61 (Oct. 1975).

Shumann, William A. "Airport, Airway Plan Hits Opposition in Congress." *Aviation Week & Space Technology* 102 (March 24, 1975).
Shumann, William A. "FAA Gets Contradictory Noise Guidance." *Aviation Week & Space Technology* 101 (August 12, 1974).
Shumann, William A. "House Unit Would Boost Airport, Airway Funding," *Aviation Week & Space Technology* 103 (September 29, 1975).
Shumann, William A. "Key Noise Reduction Decisions Imminent." *Aviation Week & Space Technology* 100 (January 7, 1974).
Shumann, William A. "Port Authority Position Clouds Concorde Operations to Kennedy." *Aviation Week & Space Technology* 102 (April 21, 1975).
Sickman, Linda M. "Fourth Amendment: Limited Luggage Seizures Valid on Reasonable Suspicion." *The Journal of Criminal Law and Criminology (1973-)* 74 (Winter, 1983).
"Silencing Jet Fleet Will Be Costly Civil Engineers Are Advised." *Aviation Week* 66 (May 27, 1957).
"$63 Million Annual Airport Aid Wins Unanimous Senate Approval." *Aviation Week* 63 (July 4, 1955).
"Stewart Airport Seeks Jet Service, But Carrier Doubt Depth of Market." *Aviation Week & Space Technology* 129 (October 31, 1988).
"Strong, Varied Opposition Meets Latest Noise Bill." *Aviation Week & Space Technology* 108 (June 5, 1978).
Sullivan, Kathleen M. and Akhil Reed Amar. "The Supreme Court, 1991 Term," *Harvard Law Review* 106 (Nov 1991).
"Surcharges Urged for Noise Compliance." *Aviation Week & Space Technology* 106 (May 9, 1977).
Swindell, David and Mark S. Rosentraub, "Who Benefits from the Presence of Professional Sports Teams? The Implications for Public Funding of Stadiums and Arenas," *Public Administration Review* 58 (Jan-Feb 1998).
Taaffe, Edward J. "Air Transportation and United States Urban Distribution." *Geographic Review* 46 (April 1956).
Taaffe, Edward J. "The Urban Hierarchy: An Air Passenger Definition." *Economic Geography* 38 (Jan., 1962).
"Task Force Fails to Agree on Airport Privatization." *Aviation Week & Space Technology* 131 (March 26, 1990).
"10-Year Aviation Facilities Plan Readied." *Aviation Week & Space Technology* 88 (January 29, 1968).
"Texas Airport May Cover 21,000 Acres." *Aviation Week* 83 (October 4, 1965).
"Text of Port Authority's 'My Deal Colleague' Letter.'" *Aviation Week* 69 (September 8, 1958).
"The Top Two Dozen—1966." *Flight International* (5 October 1967).
"T. P. Wright Announces 1947 Airport Allotments." *Airports* 11 (Feb., 1947).

"Traffic Growth Swamps Airport Facilities." *Aviation Week & Space Technology* 85 (October 31, 1966).
Trust Fund Issues Cloud Airport Air Legislation." *Aviation Week & Space Technology* 116 (March 22, 1982).
"Trust Fund Meets House Opposition." *Aviation Week & Space Technology* 112 (April 21, 1980).
"Trust Fund Raid Effort Has Little Hope." *Aviation Week & Space Technology* 102 (April 21, 1975).
"Two Airport Groups Shift Defederalization Position." *Aviation Week & Space Technology* 116 (April 5, 1982).
"Urges Airports Near Noisy Areas." *Aviation Week* 58 (January 12, 1953).
"U.S. Officials Find Concorde Noise Acceptable." *Aviation Week & Space Technology* 100 (June 24, 1974).
"The U.S. Skimps on Airport Safety," *Newsweek* XCV (March 31, 1980).
"The Validity of Airport Zoning Ordinances." *Duke Law Journal* 1965 (Autumn, 1965).
Vietor, Richard H. K. "Contrived Competition: Airline Regulation and Deregulation, 1925–1988." *The Business History Review* 64 (Spring 1990).
Voigt, Wolfgang. "From the Hippodrome to the Aerodrome, from the Air Station to the Terminal: European Airports, 1909–1945," in John Zukowski, ed., *Building for Air Travel: Architecture and Design for Commercial Aviation.* Munich and New York: Prestel-Verlag; Chicago: The Art Institute of Chicago, 1996.
"Washington Roundup: Noise Efforts." *Aviation Week & Space Technology* 104 (May 31, 1976).
Watkins, Harold D. "Airport congestion is Forcing New Wave of Expansion." *Aviation Week & Space Technology,* Vol. 83, No. 17 (October 25, 1965).
Watkins, Harold D. "Airlines' Fiscal Woes Buffet Airports." *Aviation Week & Space Technology* 95 (November 15, 1971).
Watkins, Harold D. "Miami Plans Large-Scale Training Airport," *Aviation Week & Space Technology* 89 (July 22, 1968).
Watkins, Harold D. "Traffic Sparks Airport Needs." *Aviation Week & Space Technology* 98 (May 28, 1973).
Werlich, John M. and Richard P. Krinsky, "The Aviation Noise Abatement Controversy: Magnificent Laws, Noisy Machines, and the Legal Liability Shuffle." *Loyola of Los Angeles Review* 15 (December 1981).
Wetmore, Warren C. "Concorde Suit Raises Broad Issue." *Aviation Week & Space Technology* 104 (March 22, 1976).
Wetmore, Warren C. "State Noise Bill Would Block Concorde." *Aviation Week & Space Technology* 104 (March 1, 1976).
"Where Federal Aid Cash Will Go." *Aviation Week* 53 (November 13, 1950).

Wilson, George C. "Kennedy Airport Bill Nears Critical Test." *Aviation Week* 75 (July 10, 1961).

Winston, Donald C. "Airway, Airport Formula Revised." *Aviation Week & Space Technology* 88 (June 24, 1968).

Winston, Donald C. "Congressional Conferees Restrict Aviation Trust Fund Expenditures." *Aviation Week & Space Technology* 95 (August 2, 1971).

Winston, Donald C. "Nixon Transportation Trust Fund Idea Hit." *Aviation Week & Space Technology* 94 (March 29, 1971).

Winston, Donald C. "Senate Would Halt Trust Fund Loss." *Aviation Week & Space Technology* 95 (July 26, 1971).

"White House Science Unit Seeks Government Attack on Jet Noise." *Aviation Week & Space Technology* 84 (March 28, 1966).

Whitelegg, Drew. "Keeping their eyes on the skies: Jet aviation, Delta Air Lines and the growth of Atlanta." *Journal of Transportation History* 21 (March 2000).

Yafee, Michael L. "FAA Noise Certification Seen Inevitable." *Aviation Week & Space Technology* 89 (November 25, 1968).

Yafee, Michael L. "NASA Begins Major Engine Noise Project." *Aviation Week & Space Technology* 87 (August 21, 1967).

Dissertations

Warren, Drake Edward. "The Regional Economic Effect of Commercial Passenger Service at Small Airports." University of Illinois at Urbana-Champaign, 2008

Internet Sources

"Airmall" http://www.airmall.com/(accessed June 18, 2012).

"Airport Cities Drive Local Economic Development," *MuniNetGuide* (March 23, 2010). http://www.muninetguide.com/print.php?id=361 (accessed July 13, 2010).

Airport Data (5010) & Contact Information. http://www.faa.gov/airports_airtraffic/airports/airport_safety/airportdata_5010/menu/index.cfm#reports (accessed June 17, 2009).

Airports Worldwide. http://www.airportsworldwide.com/Why-Airports-Worldwide/Who-We-Are/ (accessed December 30, 2014).

Airports Worldwide History, http://www.airportsworldwide.com/Why-Airports-Worldwide/History/ (accessed December 30, 2014).

"Annual Results US Airlines" http://airlines.org/data/annual-results-u-s-airlines-2/ (accessed June 20, 2016).

"BAA concludes sale of US airport retail management business," http://www.moodiereport.com/document.php?c_id=6&doc_id=24842 (accessed June 18, 2012)

Belson, Ken. "Delaying of Dubai Port Deal Brings Port Authority Grief," *New York Times*, February 16, 2007 http://www.nytimes.com/2007/02/16/nyregion/16dubai.html (accessed June 18, 2012).

Board of Airport Commissioners of the City of Los Angeles v. Jews for Jesus, Inc. (No. 86-104). http://www.law.cornell.edu/supct/html/historics/USSC_CR_0482_0569_ZO.html. (Accessed 12 May 2010).

Buchanan, Will. "Court upholds ban on Hare Krishna soliciting in LAX airport," *The Christian Science Monitor* March 26, 2010, http://www.csmonitor.com/layout/set/print/content/view/print/290402 (accessed March 26, 2010).

"CA Supreme Court rules on Hare Krishnas' Airport Solicitations." http://www.legalinfo.com/legal-news/ca-supreme-court-rules-on-hare-krishnas-airport-solicitations.html (accessed 12 May 2010).

Conspiracy Theories and TWA 800. : http://en.wikipedia.org/wiki/TWA_Flight_800_conspiracy_theories (accessed August 27, 2014).

"Constellation" http://aerostories.free.fr/connie/page10.html (accessed 5 August 2014).

"CY 2006 Primary and Non-Primary Commercial Service Airports (updated 10/18/2007)" http://www.faa.gov/airports_airtraffic/airports/planning_capacity/passenger_allcargo_stats/passenger/media/cy06_primary_np_comm.pdf (accessed 29 May 2008).

Dolan, Maura and DanWeikel. "L.A. can bar Hare Krisnas from panhandling at airports, court rules," *The Los Angeles Times*, March 26, 2010, http://www.latimes.com/news/local/la-me-lax-hare-krishnas26-2010mar26,0,4257614.story (accessed 12 May 2010).

Driehaus, Bob Driehaus. "DHL Cuts 9,500 Jobs in U.S., and an Ohio Town Takes the Brunt," *New York Times*, November 10, 2008, http://www.nytimes.com/2008/11/11/business/11dhl.html (accessed June 18, 2012).

EPA Press Release, "EPA To Launch Noise Control Program," November 6, 1972 http://www.epa.gov/history/topics/nca/02.html. (accessed July 4, 2011)

EPA Press Release, "EPA Proposes Quieting of Jet Airplanes," January 31, 1975 http://www.epa.gov/aboutepa/history/topics/nca/03.html (accessed July 4, 2011).

FAA Historical Chronology, 1926–1996, "September 2, 1975" http://www.faa.gov/about/media/b-chron.pdf (accessed June 19, 2012).

Federal Aviation Administration Noise Abatement Policy, November 18, 1976 http://airportnoiselaw.org/faanap-1.html (accessed July 5, 2011).

FAA Regulations. http://rgl.faa.gov/Regulatory_and_Guidance_Library/rgFAR.nsf/MainFrame?OpenFrameSet (accessed June 20, 2012)

Final call for JKF's classic TWA Terminal." http://www.cnn.com/2015/10/14/aviation/twa-terminal-jfk-airport/index.html (accessed October 23, 2015).

"Greater Cincinnati/Northern Kentucky Airports - History." http://www.cvgairport.com/about/history2.html' (accessed March 1, 2012).

Historical Air Traffic Statistics, Annual 1954–1980. http://www.rita.dot.gov/bts/sites/rita.dot.gov.bts/files/subject_areas/airline_information/air_carrier_traffic_statistics/airtraffic/annual/1954_1980.html (accessed December 30, 2014).

"History of Palmdale" http://www.lawa.org/pmd/pmdHistory.cfm (accessed 3/19/2008).

Huffenberger, Gary. "DHL to donate airpark, *News Journal*, January 18, 2010.

http://www.wnewsj.com/main.asp?SectionID=49&SubSectionID=156&ArticleID=181879 (accessed June 18, 2012).

Indianapolis Airport. http://www.indianapolisairport.com/information_news/iaa.aspx (accessed June 17, 2009).

Kadlec, Cindy. Staff Attorney, "Final Report of the Senate Interim Committee on Regional Control of Lambert-St. Louis International Airport," (February 2003), http://www.senate.mo.gov/04info/comm/interim/lambert.pdf. (Accessed August 20, 2008).

Karon, Tony with Douglas Waller. "Who's Behind the Dubai Company in U.S. Harbors?" *Time*, February 20, 2006 http://www.time.com/time/nation/article/0,8599,1161466,00.html (accessed June 18, 2012).

Los Angeles Airport. http://www.scpr.org/news/2010/07/06/observation-deck-lax-theme-building-set-re-open-pu/(accessed August 5, 2010).

Massport. http://www.massport.com/airports (accessed June 17, 2009).

Metropolitan Airports Commission. http://www.mspairport.com/mac/organization/default.apx (accessed June 17, 2009).

Metropolitan Washington Airports. http://www.metwashairports.com/about_the_authority/history. (accessed June 17, 2009).

Most Beautiful Airports. http://www.travelandleisure.com/articles/worlds-most-beautiful-airports/1 (accessed August 9, 2010).

National Transportation Safety Board Report (TWA 800). http://www.ntsb.gov/doclib/reports/2000/AAR0003.pdf (accessed August 27, 2014).

Nix, Denise. "State Supreme Court to hear Krishna LAX free speech case," DailyBreeze.com, January 1, 2010. http://www.dailybreeze.com/religion/ci_14120775 (accessed 9 March 2010).

"Observation Deck at LAX Theme building to Reopen." http://www.scpr.org/news/2010/07/06/observation-deck-lax-theme-building-set-re-open-pu/ (accessed August 5, 2010).

Orlando Sanford International Airport-Organization. http://www.orlandosanfordairport.com/organization.asp (accessed December 30, 2014).

Passenger Statistics, Lambert-St. Louis International Airport http://www.lambert-stlouis.com/flystl/media-newsroom/stats/(accessed July 21, 2011).

Pawlowski, A. "Panhandling banned at Los Angeles Airport," http://www.cnn.com/2010/TRAVEL/07/08/los.angeles.airport.solicitations/index.html?hpt=Sbin (accessed July 8, 2010).

Pittsburgh International Airport. http://www.flypittsburgh.com/ACAA_background (accessed June 17, 2009).

Port of Portland Airport. http://www.portofportland.com/PDX_home.aspx.

"Razing the I-X Center, Dick Jacobs' will and Kucinich's Integrity Now: Whatever happened to?..." http://blog.cleveland.com/metro/2010/04/razing_the_i-x_center_14_milli.html (accessed July 27, 2011).

Sanger, David E., "Under Pressure, Dubai Company Drops Port Deal," *New York Times*, March 10, 2006, http://www.nytimes.com/2006/03/10/politics/10ports.html?pagewanted=all (accessed June 18, 2012).

"707" http://history.nasa.gov/SP-468/ch13-3.htm (accessed 5 August 2014).

"707" http://www.boeing.com/boeing/commercial/707family/index.page (accessed 5 August 2014).

"747" http://www.boeing.com/boeing/commercial/747family/specs.page? (accessed 5 August 2014).

Tampa. http://www.tampaairport.com/about/administration/mission_vision.asp (accessed June 17, 2009).

Teaher, David. "Ferrovial lands BAA with final offer of £10.bn," http://www.guardian.co.uk/business/2006/jun/07/theairlineindustry.travelnews (accessed June 26, 2009).

"TSA Frequently Asked Questions – Program" http://www.tsa.gov/what_we_do/optout/spp_faqs.shtm (accessed July 8, 2010).

"TSA – Our History." http://www.tsa.gov/research/tribute/history.shtm (accessed July 9, 2010).

TSA Screening Partnership Program. http://www.tsa.gov/stakeholders/screening-partnership-program (accessed August 23, 2014).

US Travel Association, "Travel Facts and Statistics," http://www.ustravel.org/news/press-kit/travel-facts-and-statistics (access August 1, 2013).

Weikel, Dan. "Palmdale City Council moves to develop shuttered airport," *Los Angeles Times*, April 3, 2009 http://www.latimes.com/wireless/avantgo/la-me-palmdale3-2009apr03,0,877428.story (accessed April 3, 2009).

"Why CVG Lost Half All Flights." http://news.cincinnati.com/article/20100524/EDIT03/5230393/Why-CVG-lost-half-all-flights (accessed March 1, 2012).

Other

Ballal, Dilip R. and Joseph Selina, "Progress in Aero Engine Technology (1939–2003)" in "Centennial of Powered Flight Celebrations," Offprint of Special Issue on Air Transportation, *AIAA Journal of Aircraft*, 2003.

Discazeaux, Carine and Mario Polese. "Cities as Air Transport Centres: An Analysis of the Determinants of Air Traffic Volume for North American Urban Areas." Montreal: Institut national de la rechereche scientificque Urbanisation, Culture et Societe, working paper, November 2007.

Index

J.R. Bednarek, *Airports, Cities, and the Jet Age*, DOI 10.1007/978-3-319-31195-1